Computational Analysis of Terrorist Groups: Lashkar-e-Taiba

V. S. Subrahmanian · Aaron Mannes
Amy Sliva · Jana Shakarian
John P. Dickerson

Computational Analysis of Terrorist Groups: Lashkar-e-Taiba

Foreword by R. James Woolsey

 Springer

V. S. Subrahmanian
Institute for Advanced Computer Studies
University of Maryland
College Park, MD
USA

Aaron Mannes
Institute for Advanced Computer Studies
University of Maryland
College Park, MD
USA

Amy Sliva
College of Computer and Information
  Science
Northeastern University
Boston, MA
USA

Jana Shakarian
Institute for Advanced Computer Studies
University of Maryland
College Park, MD
USA

John P. Dickerson
Department of Computer Science
Carnegie Mellon University
Pittsburgh, PA
USA

ISBN 978-1-4899-9263-5   ISBN 978-1-4614-4769-6 (eBook)
DOI 10.1007/978-1-4614-4769-6
Springer New York Heidelberg Dordrecht London

Foreword

Aaron Mannes and V. S. Subrahmanian have produced a fascinating framework for a disciplined analysis of terrorist groups.

Using as their case the Pakistani-backed Lashkar-e-Taiba ("LeT"), the perpetrator of the dramatic attack on Mumbai in November 2008, the authors examine in detail the group's ideology, history, and all pertinent facts available, from their relations with other Islamist groups to their propensity for attacks on holidays.

The authors' objective is to derive from the data what they term "temporal probabilistic" rules to determine what actions to thwart LeT's campaign will likely have the most success and indeed which of our actions may effectively counter LeT attacks of one type yet actually weaken us against others.

The authors' care and precision are exemplary and yield a rich beginning to the potential utilization of such probabilistic modeling as a broadly used tool for combatting many types of terrorism.

April 17, 2012

R. James Woolsey

Preface

Computation is being used in almost every discipline today—most of the sciences and engineering view computation as an indispensable tool in their efforts to understand one or more phenomena.

This book presents the first in-depth and comprehensive study of a real-world terrorist group using the same kind of "big data" analytic methodologies that have enabled companies like Google and Amazon to model the behaviors of customers and users of their Web services.

We chose Lashkar-e-Taiba as the subject for this rigorous computational analysis in early 2007. Although LeT has been around for over 20 years, outside of terrorism and South Asia specialists, LeT was only taken seriously in the west after the November 26, 2008 attacks in Mumbai.

We chose to conduct this study with a mix of computational science, social science, and public policy researchers, so that methodologies from these diverse disciplines would jointly inform our understanding of Lashkar-e-Taiba's behavior and enable us to shape policies towards them.

We thank many people for their assistance with this work. First, we thank Stephen Tankel—the author of the first (excellent) book on LeT—for reading previous drafts and providing detailed comments. His insights and comments were invaluable. Second, we thank Animesh Roul also for going through the manuscript in detail, providing numerous corrections and references that we had overlooked.

On the technology side, we thank several people who worked on versions of the technology that were eventually used in this book. TP-rules were invented by V. S. Subrahmanian, together with his former Ph.D. student, Alex Dekhtyar. Algorithms to learn TP-rules automatically from data were developed by V. S. Subrahmanian and his former student, Jason Ernst. They form the technical backbone of much of this book. The use of mixed integer linear programming for generating policies was based on work by V. S. Subrahmanian with his then Ph.D. student Raymond Ng, Anil Nerode at Cornell, and Colin Bell at Iowa. We thank Damon Earp for setting up the database system through which our data were collected and stored. Dan LaRocque and LTG (Ret.) Charley Otstott also helped build systems to explore LeT's network. Roy Lindelauf at the Netherlands

National Defense Academy (NLDA) has also been a valuable sounding board for terrorism-related studies.

All of the research work done on this book was done when the authors were at the University of Maryland. We are grateful to UMIACS for their technical support of our work. We are also very grateful to sponsors who funded some of the antecedents of the technologies used in this study. The Air Force Office of Scientific Research initiated funding of our work on computational models of terrorism via grants FA95500610405 and FA95500510298. A policy analytics framework, as well as the TP-rules used in this work (beyond the original TP-rule framework from the 1990s), were studied through generous funding from the US Army Research Office under grant W911NF0910206. In particular, we would like to thank Dr. John Tangney at ONR who got our work in this field started, Dr. Terry Lyons at AFOSR (now sadly deceased) for his strong support of our research, as well as Dr. Purush Iyer at ARO for encouraging us to conduct both theoretical and applied research in this emerging field. Last, but not least, we thank Barbara Lewis and Jennifer Newlin for helping typeset the book in the correct format.

College Park, April 2012

Contents

Acronyms

APT	Annotated Probabilistic Temporal
ASIO	Australian Security Intelligence Organization
BBC	British Broadcasting Corporation
BSF	Border Security Force
CARA	Cultural Adversarial Reasoning Architecture
CIA	Central Intelligence Agency
CII	Council of Islamic Ideology
CRPF	Central Reserve Police Force
DPC	Defense of Pakistan Council
DST	Direction de la Surveillance du Territoire
FARC	Fuerzas Armadas Revolucionaries de Colombia
FDLR	Forces Democratiques des Liberation du Rwanda
FIF	Falah-i-Insaniyat Foundation
FTO	Foreign Terrorist Organization
HM	Hizb-ul-Mujihideen
IED	Improvised Explosive Device
IKK	Idara Khidmat-e-Khalq
IM	Indian Mujihideen
J&K	Jammu and Kashmir
JeM	Jaish-e-Mohammed
JKLF	Jammu & Kashmir Liberation Front
JuD	Jamaat ud-Dawa
KGB	Soviet Committee for State Security
KN	Khairun Naas
IMF	International Monetary Fund
ISI	Inter Services Intelligence
LeT	Lashkar-e-Taiba
LoC	Line of Control
MDI	Markaz al-Dawa Irshad
MEMRI	Middle East Media Research Institute
MJAH	Markazi Jamiat Ahl Hadith

NCTC	National Counterterrorism Center
OGW	Over Ground Workers
PCA	Policy Computation Algorithm
POK	Pakistan Occupied Kashmir
PTI	Press Trust of India
SATP	South Asian Terrorism Portal
SDGT	Specially Designated Global Terrorist
SIMI	Student Islamic Movement of India
SOMA	Stochastic Opponent Modeling Agent
TIM	Tanzim-Islahul-Muslimeen
TP	Temporal Probabilistic
USA	United States of America
VoIP	Voice over Internet Protocol
WITS	Worldwide Incident Tracking System

Chapter 1
Introduction

Abstract This chapter contains a brief introduction to Lashkar-e-Taiba. It describes their geographic locations and also summarizes statistics about violent terror acts carried out by LeT, briefly describes the behavioral rules about LeT derived in this book, and summarizes suggested policy options generated automatically from the LeT data set.

For three full days starting on November 26, 2008, the eyes of the world were on Mumbai as ten gunmen held the city under siege. The sites varied dramatically in scope—from a crowded train station used daily by hundreds of thousands of commuters from every segment of Indian society, to a pair of exclusive hotels frequented by the wealthiest Indians and foreign visitors, to a little known Jewish center on a narrow street (Fair 2009; John 2011). Through a carefully calculated and meticulously planned operation, the terrorists were able to hold Indian police and counter-terror forces at bay for over 72 hours, while simultaneously commanding continuous coverage from the world's press. Later investigations revealed that the terrorists involved used sophisticated new technologies including Global Positioning System (GPS) and voice over IP (VoIP) technologies to execute their attack and communicate with their handlers back in Pakistan (Government of India 2008).[1]

Who were these terrorists? Which group did they belong to and how did this group get its power? How were they recruited? Where did they get the training required to hold off a nuclear power's armed forces several days? Who provided them with the political, financial, and military support needed to execute such a well-coordinated and devastating attack? In the shadowy world of inter-national terrorism, answers to such questions are not always apparent. In this case, however, Indian police captured one of the terrorists, Ajmal Kasab, alive. After

---

[1] This report was released by the Government of India to the Government of Pakistan. Page 2 of the report de-scribes the retrieval of GPS instruments and a Thuraya satellite phone. Page 12 of the same report describes the use of VoIP technology by the terrorists.

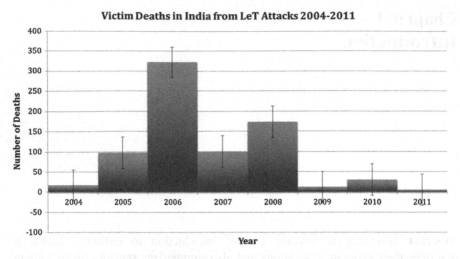

Fig. 1.1 Total number of people killed in India by LeT or its Indian allies based on data from the National Counterterrorism Center's Worldwide Incident Tracking System (WITS) database. There is some question about LeT's responsibility for the attacks in India that have been attributed to it, particularly the 2006 Mumbai train bombings that killed over 200 people. The WITS website discusses its criteria for identifying the perpetrators of terror attacks here: http://www.nctc.gov/witsbanner/wits_subpage_criteria.html

extensive interrogation, it emerged that the group behind the attacks was Lashkar-e-Taiba (LeT for short), the "Army of the Pure (or Pious)[2]" (Government of India 2008, p. 2 point 9). Subsequent testimony in a Chicago courtroom by David Headley, a US citizen of Pakistani origin, who conducted operational surveillance of the locations targeted in the Mumbai assault, confirms not only the involvement of Lashkar-e-Taiba in the Mumbai attacks, but also the fact that Pakistan's Inter Services Intelligence (ISI) agency may have been involved (Rotella 2011, also *US Government vs. Tahawwur Rana* transcripts).

Well-known to specialists on terrorism and south Asian affairs, LeT has been responsible for many terrorist attacks in India, Kashmir, Pakistan, and Afghanistan. In addition, LeT operative Faheem Lodhi was arrested and convicted (in 2006) of planning sophisticated attacks on Australia's power grid (Brenner 2011; The Age 2006) demonstrating a potential global threat.

Figure 1.1 shows that hundreds of people have been killed in attacks attributed to LeT and its allies over the past 15 years based on data from the US National Counterterrorism Center Worldwide Incident Tracking System (WITS) database.

The red bars show the number of people killed during a given year—the black error lines overlaid on the red bars show margins of error. For instance, according to Fig. 1.1, we believe about 325 people (shown by the red bar) were killed in

---

[2] It has also been asserted that the name Lashkar-e-Taiba should be read as "Army of Madinah" (John 2011).

India by LeT in 2006, but the true number (shown by the black error line) may be anywhere between 280 and 370. Uncertainty arises in the numbers shown because it is not always clear whether LeT was responsible for a given attack.

Long used as a proxy by Pakistan's military in its ongoing confrontation with India in Kashmir (Tellis 2010), LeT had established a proven ability to carry out attacks across the Indian sub-continent—primarily in India, Kashmir, and (as the conflict in Kashmir has calmed) with increasing reports of LeT involvement in Afghanistan (Rubin 2010). LeT has also been able to mount spectacular attacks including the November 26, 2008 Mumbai attacks that left 166 innocent people dead[3] (plus an additional 9 terrorists), and a December 2000 attack on soldiers at the Delhi historic site Red Fort. The Indian government also held LeT responsible for the July 2006 Mumbai train bombings that left approximately 211 people dead; however there is some dispute about LeT's role in the attack.[4]

The ability of LeT to hold a major city in terror for several days and exacerbate tensions between a pair of frequently sparring nuclear powers places an accurate understanding of this group as a major priority for those concerned with international security affairs (Krepon 2010). Beyond its formidable paramilitary capacity, LeT runs social service networks and businesses in Pakistan ensuring a steady stream of new recruits and increasing the number of adherents to its radical jihadist ideology, potentially contributing to Pakistan's radicalization (Abbas 2005). Finally, LeT is showing the potential for carrying out international attacks, making it a possible threat to the Western interests it excoriates rhetorically (Tankel 2011a). LeT's growing power indicates that gathering and analyzing information about LeT needs to be done sooner, rather than later.

Who are the LeT? Who founded them and why? What are their principal goals? What are their principal grievances and what actions have they taken to address them? Who are their principal supporters? What is their modus operandi? Who do they target—and under what conditions? More importantly, what might LeT do in the future and how can the various affected parties act in order to minimize the damage they cause not only to innocent civilians, but also to peace and stability in South Asia?

This book is an attempt to answer some of these questions definitively on the basis of a principled analysis of data, and to provide suggestions that may help national security policy makers and military leaders define a strategy to combat LeT's growing influence and capabilities.

---

[3] In all casualty figures quoted in this book, the number of terrorists killed is not included. Thus, when the number of casualties for the November 26, 2008 Mumbai attack is listed as 166, it does not include the terrorists who were killed during the attack.

[4] The other major suspect in the attack was Indian Mujahideen (Gupta 2011), an Indian Muslim terrorist group that receives support from LeT (this relationship is discussed below). Because of the frequent ambiguity in identifying the perpetrator of a terrorist attack, this book will use the term LeT-backed attacks to include attacks that are believed to have been carried out either directly by LeT or by its proxies and close allies.

Unlike many books in the realm of public policy, the recommendations made in this book are through a mix of two very different analytical techniques—traditional qualitative policy analysis and a detailed, mathematical and computational analysis of systematically gathered and carefully curated data about LeT's behavior since its inception in the mid-1980s.[5] The computational analysis, though certainly not infallible, provides confidence that behavioral cues have not been missed and that the space of possible policies that may be used to influence their behavior has been thoroughly investigated. The computational analysis is based on the CARA (Cultural Adversarial Reasoning Architecture) platform (Subrahmanian et al. 2007; Subrahmanian 2007) which brought "big data" mining and analytics techniques to the analysis of terrorism data for the first time. Nonetheless, readers should keep in mind that computational methods are also not guaranteed to be exhaustive or even fully correct as they are only as "good" as the data they use. Yet, computational approaches enabled the mathematical analysis of over 770 variables relating to LeT with data on a monthly basis stretching back to January 1990—something that any human might find very challenging to do.

This book is organized as follows. The remainder of this chapter will first provide a broad overview of LeT's activities containing summary statistics about their behavior. This chapter also summarizes the study's findings about LeT's organizational behavior and policies that might be used to counter their growing influence.

Chapter 2 describes the creation of LeT, its ideology and organization, the ecosystem of governmental, religious, criminal, and terrorist organizations that support LeT, and an overview of its major theaters of operations—Pakistan, Jammu & Kashmir,[6] the rest of the India sub-continent, and the international arena. *This book principally focuses on LeT's operations in India and Pakistan (including the portions of Kashmir controlled by each country).* Reliable data on LeT operations in other geographies (particularly Afghanistan where LeT have expanded operations since the mid-2000s) is limited and hard to analyze.

Chapter 3 discusses the principal technical device used in this book—temporal probabilistic rules (or TP-rules for short). TP-rules are stochastic rules of the form

---

[5] The precise date on which LeT was founded is not 100 % clear (Tankel 2011b, p. 3) traces the roots of LeT back to 1984 when Zaki-ur-Rehman Lakhvi, one of the principal suspected masterminds behind the 2008 Mumbai attacks, founded an Ahl-e-Hadith group in Afghanistan to fight the Soviets. Tankel goes on to report that Hafez Saeed (the current LeT chief) and Zafar Iqbal, created Jamaat-ud-Dawa (JuD), an Ahl Hadith missionary group in 1985. Tankel describes the subsequent creation of Markaz-ud-Dawa al Irshad (MDI) in 1986 by 17 founders including those above. The new organization folded in Zaki-ur-Rehman's group and Saeed's organization. LeT itself was officially launched in 1990, though there appear to be conflicting reports about the exact date (John 2011, p. 1) states that LeT was created precisely on February 22, 1990.

[6] Throughout this book, the term Jammu & Kashmir is used to refer to Indian administered Kashmir. This expression is used for convenience, the issue of Kashmir's ultimate legal status is a complex one and is not the topic addressed in this book.

If the environment in which LeT operates satisfies some condition C during a given month, then there is a probability of P % that LeT will take action A at intensity level i, t months later.

Thus, TP-rules allow an analyst to look at the situation today and predict what LeT might do 1, 2, 3 or more months (i.e., t can be 1, 2, 3, or more) later. Individual TP-rules cannot be singled out conveniently and used piecemeal to make such predictions—rather, the entire set of rules must be used in conjunction as many TP-rules are related to other TP-rules.

Chapters 4, 5, 6, 7, 8 and 9 describe specific kinds of attacks carried out by LeT. Chapter 4 examines the conditions (and hence identifies TP-rules) associated with attacks on civilians (largely Hindu) by LeT. Chapter 5 describes TP-rules describing the conditions under which LeT launches attacks against public sites, tourist sites, and transportation networks (such as the July 2006 Mumbai train bombings). Chapter 6 describes TP-rules about attacks carried out by LeT against professional security forces such as the Indian Army. Chapter 7 describes TP-rules pertaining to attacks by LeT against security installations and infrastructure such as Army bases. Chapter 8 provides a description of TP-rules pertaining to other types of attacks: specifically attacks carried out on holidays, attacks against government sites, and attempted (but unsuccessful attacks). Chapter 9 examines TP-rules derived about armed clashes involving LeT—armed clashes are violent encounters between LeT and government security forces in which LeT did not possess the initiative.

Chapter 10 briefly describes the mathematical and computational "policy analytics" techniques used to generate policies from the TP-rules described in Chaps. 4, 5, 6, 7, 8 and 9. In particular, Chap. 10 proves a critical theorem showing that it is impossible (at least theoretically) to fully eliminate all of LeT's violent actions. But it does provide the theoretical framework under which most of the types of attacks carried out by LeT can be reduced. Chapter 11 uses the mathematical and computational model defined in Chap. 10 to suggest policies that might be useful in mitigating most of LeT's violent behavior. *These policies are not sufficient to guarantee "good" behavior from LeT*—rather, it would appear that these policies would significantly mitigate risk of violence from LeT, though it is unlikely that they will eliminate all such bad behavior.

Finally, this book includes a set of Appendices.

- Appendix A describes the data methodology used in this research including information on the type of data collected, the data sources consulted, and the types of curation performed.
- Appendix B shows a summary of all terrorist attacks attributed to LeT by the US National Counterterrorism Center's Worldwide Incident Tracking System.
- Appendix C shows all the TP-rules discussed in this paper.
- Appendix D shows all the eight policies derived by the Policy Computation Algorithm—four of which are discussed in detail in Chap. 10 (the other four are very similar and only differ on minor points).

- Appendix E summarizes reports of any kind of internal strife or intra-organizational dissension within LeT. Though LeT experienced only one major split, reports of internal dissension have occurred more often.
- Appendix F summarizes reports documenting LeT's relationship with the Pakistani military.
- Appendix G summarizes reports documenting LeT's relationship with the press.
- Appendix H summarizes information derived from open sources on LeT's locations including training camps.

All the data used in this study was gathered from *open sources*—primarily news sources, but also reports, academic papers, and books.

1.1 How to Read this Book

With the exception of Chaps. 3 and 10, all chapters of this book should be accessible to all readers. A qualitative social scientist, law enforcement or intelligence official, or a policy expert, may want to skip Chaps. 3 and 10 unless they are specifically interested in the details of the "data mining" used to generate the rules about LeT behavior. Nonetheless, Chaps. 3 and 10 should be somewhat accessible to the layperson as it is presented at a high level without gory mathematical technical details.

In addition, all chapters describing TP-rules (Chaps. 4, 5, 6, 7, 8 and 9) contain a summary figure. While this figure is not a substitute for the detailed presentation contained within these chapters, it does provide a birdseye view of some relationships between variables describing the environment in which LeT operates, and the actions it takes. A reader can quickly peruse this figure to get a general idea of the relationship between LeT's violent acts and the possible triggers.

1.2 LeT's Camp Locations

This section briefly describes the locations of LeT's major training camps and offices. LeT's main camp is in Muridke, just outside the city of Lahore, in the Pakistani state of Punjab. More details on training camps are listed in Appendix H.

Figure 1.2 shows locations of known LeT training camps using data collected from open sources. Not every camp is plotted on the map—if there are multiple facilities somewhere, just a single blue icon denotes the existence of camps at that location.

Figure 1.2 is obviously incomplete and is restricted to training camps and major offices that carry out LeT activities. It does not include the locations of LeT's dozens of district offices or the thousands of locations that host "charity boxes" for collecting funds for LeT activities (Rana 2006). Figures 1.3 and 1.4 break these

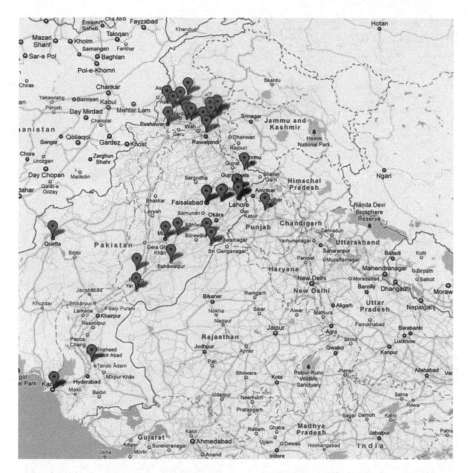

Fig. 1.2 Map showing locations of LeT training camps/offices as of December 31, 2010

camps down into those in Pakistan versus those in Jammu and Kashmir (most camps are in the so-called "Azad Kashmir" region of Pakistan currently controlled by the Pakistani Army).

1.3 Summary Statistics About LeT's Violent Activities

This section provides statistics about violent activities by LeT and its allies. The data used to generate these statistics was from the US Government's National Counterterrorism Center (NCTC) Worldwide Incident Tracking System (or WITS) database which represents not only the US government's definitive listing of terrorist acts (a violent act needs to satisfy various legal requirements for it to be considered a "terrorist" incident according to US law), but also the US

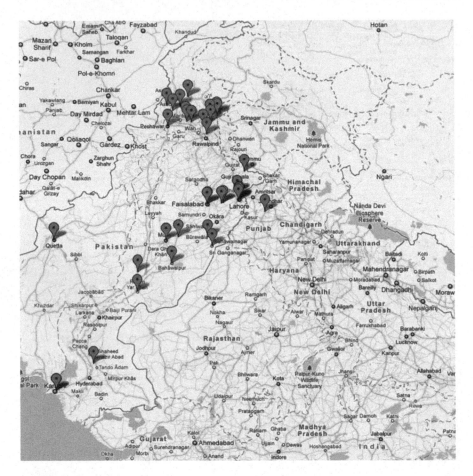

Fig. 1.3 LeT camps in Pakistan

government's determination of the perpetrators behind the act (United States Government, National Counterterrorism Center, Worldwide Incidents Tracking System 2011). Though there are many databases of terrorist incidents, WITS is used to generate these statistical summaries as it has the full weight and authority of the US government behind it. We note that the WITS data does not always "pin" responsibility for an attack on a group—but may make statements less definitive. All attacks where WITS' data implied that LeT may have done it were attributed to LeT.

Figure 1.5 shows the number of attacks carried out by LeT and its allies on an annual basis during the 2004–2011 period. The data for 2011 is incomplete—though it was compiled in late-December 2011.

Figure 1.6 shows attacks in Pakistan that were attributed to LeT by the NCTC.

Figure 1.7 shows the number of attacks attributed to LeT by the NCTC that occurred in India during the same period.

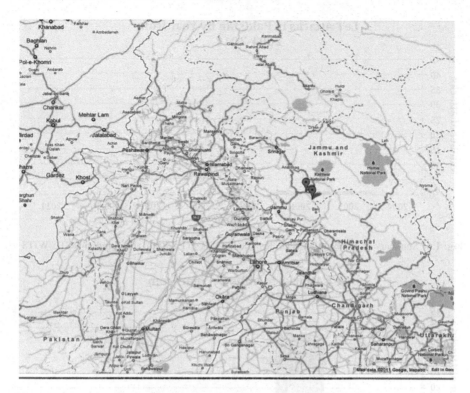

Fig. 1.4 LeT camp locations in Jammu & Kashmir

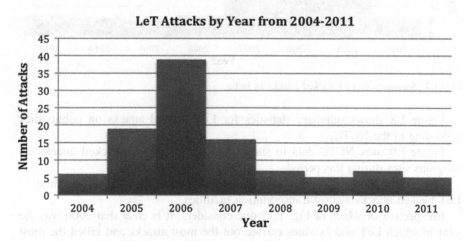

Fig. 1.5 Summary of number of attacks carried out by LeT by year

**LeT Attacks in Pakistan by Year from
2004-2011**

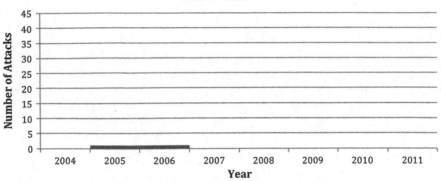

Fig. 1.6 Summary of number of attacks attributed to LeT in Pakistan, according to NCTC WITS

**LeT Attacks in India by Year from
2004-2011**

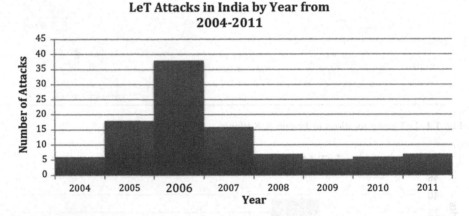

Fig. 1.7 Summary of LeT backed attacks in India

Figure 1.8 shows summary statistics for LeT backed attacks on public sites according to the NCTC.

Figure 1.9 uses NCTC data to show statistics about LeT backed attacks on religious sites during this period.

Figure 1.10 shows summary statistics generated from NCTC data pertaining to LeT backed attacks against transportation facilities.

Irrespective of which of Fig. 1.10 one considers, it is clear that 2006 was the year in which LeT and its allies carried out the most attacks and killed the most people. However, as mentioned earlier, one of the major cause of fatalities in 2006 was the July 2006 Mumbai blasts which has been variously attributed to different organizations (e.g., Indian Mujahideen).

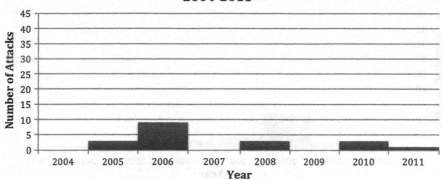

Fig. 1.8 Summary statistics on LeT backed attacks against public sites

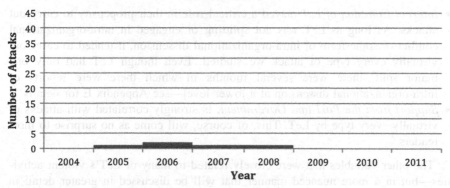

Fig. 1.9 LeT backed attacks against religious sites

1.4 Summary of Significant TP-Rules

Using the data gathered entirely from open sources (Appendix A contains details of the methodology), the system derived over 15,000 TP-rules associated with a wide variety of LeT's violent actions. These rules were painstakingly analyzed to identify the ones that satisfied two conditions:

- Quantitative conditions requiring, informally, that the rules be both accurate enough and be valid often enough—these are measured through conditions called *probability, support, inverse probability, and negative probability* and
- Qualitative conditions requiring, informally, that the rules could be explained to a researcher or policy maker without a quantitative background in "plain" English.

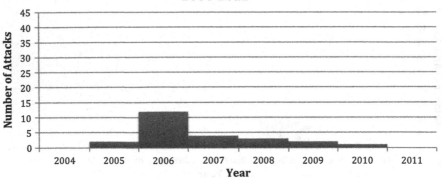

Fig. 1.10 Attacks by LeT and its allies against transportation

Table 1.1 summarizes the findings.
Specifically, rules that fulfilled the study's conditions indicated that:

- *Internal Cohesion of LeT* played a critical role in their propensity to carry out attacks. As long as LeT was not splitting or engaged in intra-organizational conflict or some form of intra-organizational dissension, it tended to carry out virtually every type of attack we studied. Even though LeT had only one major split, there were several months in which there were reports of intra-organizational dissension at a lower level—see Appendix E for details.
- *Support from the Pakistani Government*, is strongly correlated with attacks of virtually every type by LeT. This, of course, will come as no surprise to most readers.[7]

The other variables that were closely related to many of LeT's violent activities—but in a more nuanced manner that will be discussed in greater detail in Chaps. 4, 5, 6, 7, 8 and 9—include the following:

- *Deaths of LeT commanders* seems to be positively correlated with some subsequent actions (in the next 1–3 months) by LeT such as attacks on Indian civilians and armed clashes in Jammu & Kashmir, but negatively correlated with all the other kinds of attacks studied.
- *Government (typically Pakistani) Action against LeT* (which may, in some cases, lead to the deaths of LeT commanders referenced above) is also positively correlated with some subsequent actions (in the next 1–3 months) by LeT such as attacks on civilians and armed clashes.

---

[7] Some of these variables such as Pakistani military support for LeT, LeT's communications campaigns, and charitable activities are effectively constant conditions. In this study, these conditions were coded when a major media source mentioned them as occurring. The fact that major media cites this activity could indicate a trend that this activity was occurring at a greater level (or at a more noticeable level) than at other times.

Table 1.1 Summary of relationship between variables

Attack type ↓ Variable	Death of leaders	Government action	Internal cohesion	Desertion	Release of arrestees	Rel w/Pakistan government	Comm. campaign	Charitable acts	Trial/ tribunal
Civilians	Yes	Yes	Yes	Yes					
Public sites	No	No	Yes		No				
Professional security forces		?	Yes	No	Yes	Yes	Yes		
Security installations	No	?	Yes			Yes			
Holidays	No	No				Yes			
Government			Yes		Yes	Yes			
Attempted attacks	No	?	Yes	Yes		Yes			
Armed clashes	Yes	Yes	Yes			Yes		Yes	Yes

A "yes" indicates that the variable is correlated (positively) with a particular type of "attack" while a "no" indicates a negative correlation. A "?" indicates conflicting evidence

- *Desertion by LeT members* seems to be followed by subsequent LeT attacks against Indian civilians, but there is no evidence for any other type of attack.
- *Release of (LeT) Arrestees* seems to be followed in subsequent months by attacks against Indian security forces.
- *LeT's ongoing Communications Campaigns* using news media and periodicals to get an LeT-backed message out to the public seem to be followed, within 1–3 months, by attacks on (usually Indian) security installations. Appendix G contains information about LeT's relationship with the press.
- *LeT's Charitable Acts* seems to be followed, within 1–3 months, by armed clashes with (usually Indian) security forces.
- *International Trials and Tribunals of LeT personnel* also seem to be followed, within 1–3 months, by armed clashes with (usually Indian) security forces.

The above description provides a succinct summary of some of the findings described in this book. However, this summary does not do justice to the findings and the actual TP-rules discovered are themselves much more nuanced and described in more detail in the rest of this book.

1.5 Summary of Policy Recommendations

Informally speaking, a policy is a set of do's and don't's (e.g., do this, but do not do that). However, not doing certain things can be just as important as doing others.

Chapter 10 of this book formally defines a *policy* against LeT as a set of actions (positive or negative) that an anti-LeT country or agency (e.g., India, the USA, or the CIA) can either take or not take that has the effect—as predicted by the TP-rules described in Chaps. 4, 5, 6, 7, 8 and 9 modeling LeT (that were automatically learned from historical data)—of creating circumstances that are likely (but not guaranteed) to stop *all but one of* the kinds of terror attacks that are studied in Chaps. 4, 5, 6, 7, 8 and 9 of this book.[8] Specifically, Chaps. 4, 5, 6, 7, 8 and 9, study:

- Attacks against civilians (mostly Hindus)
- Attacks against professional security forces
- Attacks against security installations
- Attacks against government and/or public sites
- Attacks against tourist sites
- Attacks on transportation infrastructure
- Attempted (but failed attacks)
- Attacks on holidays.

---

[8] Attacks on holidays are not handled directly by the policies generated automatically for technical reasons discussed in Chaps. 10 and 11. However, Chap. 11 shows that the policies recommended also have a strong effect in reducing attacks on holidays.

Table 1.2 Summary of 7 actions to be performed and 5 actions not to be performed that are common in all 4 policies generated automatically by the Policy Computation Algorithm

Actions to be performed in all policies	Actions not effective in all policies
Promote splintering of LeT	No government raids on LeT personnel
Promote intra-organizational conflict within LeT	No government arrests of LeT personnel
Disrupt Pakistani military support for LeT	No government arrest warrants for LeT personnel
Disrupt LeT's communication campaigns	No LeT personnel killed by government
Disrupt releases of arrested LeT personnel by Pakistani government	No mechanisms to encourage defection by LeT members
Encourage resignation of LeT leaders	
Disrupt LeT training camps	

Chapter 10 briefly describes a mathematical technique to generate policies associated with rules, based on the concept of "minimal models" of logic programs (Minker 1982) and a sophisticated integer programming based algorithm to compute all such minimal models that greatly leverages a prior algorithm by (Bell et al. 1994a, b). The resulting algorithm is called the Policy Computation Algorithm (PCA for short).

A major theorem proved in Chap. 10 asserts that *there does not exist a policy that is likely to prevent all the types of terrorist attacks carried out by LeT.* Though this is a mathematically proven theorem, the situation is not as dire as a straightforward interpretation of the theorem might suggest.

Further investigation, described in Chap. 10, reveals that *maximal consistent subsets* (Baral et al. 1992) of the bodies of the TP-rules described in Chaps. 4, 5, 6, 7, 8 and 9 may be considered instead. They reveal that policies can be generated that would significantly reduce the probability of *all* the types of LeT-backed terror attacks listed above with the sole exception of LeT attacks on holidays. Table 11.3 in Chap. 11 actually shows that there is a high probability that even attacks on holidays can be reduced via the policies generated by us.

The Policy Computation Algorithm automatically generates a total of eight policies, which are summarized in Appendix D. However, four of these eight policies are almost identical to the other 4, and hence the book discusses four policies that we call P1, P2, P3, and P4.

There are two factors that are remarkable about the policies discovered by the PCA.

First, the policies generated by PCA are *complex*, consisting of 12 actions that are *common* to all four policies. These 12 actions include seven do's and five don't's and are listed in Table 1.2. In other words, all four policies unanimously agreed that 12 things needed to be done (or not done).

In addition, each of the four policies P1, P2, P3, and P4 required the performance of two more actions each and here too, there was considerable overlap between the policies.

- Disrupting support provided by LeT to other Islamist organizations was required in both policies P1 and P3.
- Ensuring that there is a continuing ban on LeT by the Pakistani government was required by both policies P1 and P2.
- Ensuring that publicity for high-profile trials of LeT personnel (especially those in Australia) is minimized was required by both policies P2 and P4.
- Disrupting LeT's ability to deliver social services and medical programs was required by policies P3 and P4. Much like Hamas on the West Bank, and Hezbollah in Lebanon, LeT provides social services that enable it to command the support of segments of the Pakistani population.

In short, the Policy Computation Algorithm recommended taking a total of *sixteen* actions in combination.

While the details of these recommendations and the rationale behind them are described in detail in Chap. 11—a brief summary is presented below.

Observation 1. There is a very strong relationship between months when LeT is internally cohesive (i.e., it is not splintering or involved in intra-organizational conflict or when there are reports of other forms of internal dissension) and the occurrence of various kinds of attacks 1–3 months afterwards. Conversely, when LeT encounters internal dissension, the probability that they will (in the next 1–3 months) carry out attacks against public sites, attacks against symbolic/tourist sites (like the December 23, 2000 attack on Delhi's Red Fort), attacks on transportation hubs (such as the January 2001 attack on Srinagar Airport), attacks against civilians, attacks against the Government, or attempted (but unsuccessful) attacks is relatively low. However, even in the case of months when LeT is engaged in intra-organizational dissension, there is still a high likelihood of attacks on professional security forces, attacks on security installations, and armed clashes 1–3 months afterwards. *The first proposed policy is to disrupt internal cohesion of LeT via appropriate covert action.* Chapter 11 provides some suggested tactical methods of covert action though these are far from exhaustive. For instance, the recent announcement by the US Government of a \$10 M bounty for LeT chief Hafez Saeed (BBC, April 2012) may be viewed in two ways—first, as a device to increase the level of paranoia within LeT's leadership circles—decreasing internal cohesion through lack of trust in others—and second, as a genuine hope that some Pakistani individuals will find a way to turn in Hafez Saeed. A separate study by some of the authors (Dickerson et al. 2011) makes similar recommendations (for additional covert action against LeT) based on a game-theoretic analysis.

Observation 2. It will come as no surprise to most readers that LeT's relationship with the Pakistani Government (with special emphasis on the Pakistani military including its powerful Inter Services Intelligence agency) plays a significant role in LeT attacks. The collected data shows that when the Pakistani Government (including the military and ISI) have reduced their support for LeT, there is a strong reduction in the number of attacks on professional security forces, on security installations, and on armed clashes—these are the kinds of attacks not covered by the previous rules. Appendix F documents some of the data documenting Pakistani military support to LeT and the intensity of support was gleaned from such reports.

The second major policy recommendation is to *disrupt* the relationship between the LeT and the Pakistani government. The past 20 years have seen a single, recurring pattern—LeT carries out a major attack (usually in India), India protests, Pakistan arrests (or house arrests) some LeT leaders and throws lower level LeT operatives in jail, only to release them a few months later. *There have been no serious consequences for Pakistan (as a nation) or for the Pakistani government for LeT's terrorist acts.*

Disrupting the relationship requires the development of a credible deterrent for LeT and its allies in the Pakistani military. As suggested earlier in the study of conflict by Nobel Prize winner Tom Schelling (Schelling 1980) such a deterrent needs the form of a *public threat* by a foreign government that if LeT carries out certain types of attacks, then Pakistan can expect certain kinds of non-violent retaliation. The public threat should be made in a manner that makes it clear to the Pakistani military that the party making the threat cannot easily "back away" from the threat if the conditions triggering the threat occur. Because Pakistan is a nuclear power, India must carefully design its threats to be credible but not risk an escalation. The United States also has policy challenges with Pakistan and must ensure that any leverage it seeks to use on Pakistan will not threaten its other interests. Such retaliatory steps could include reduction or elimination of foreign aid from some international donors and/or multilateral development banks, threats to reduce water flow into Pakistan from the Indus River (India controls the head waters of some of the major tributaries of the Indus, Pakistan's major source of fresh water),[9] import restrictions on Pakistani textiles, and other economic steps. However, literature on deterrence and the related concept of coercive diplomacy (in which threats and limited force are used to persuade an actor to cease a particular behavior) requires that these threats be credible (Craig and George 1990).

Observation 3 .Disrupting LeT's relationship with the news media and peri-odicals. The TP-rules automatically derived from the data using data mining algorithms shows that when LeT mounts a publicity campaign using the news media and periodicals to get their message out, there is a strong possibility that this will be followed within 1–3 months by attacks on professional security forces, security installations, and on armed clashes. Appendix G documents some of the data on publicity efforts by LeT. *The third recommendation is that anti-LeT entities should attempt to disrupt LeT's relationship with the news media and periodicals.* There are many tactics that can be used to achieve this aim—while this study does not recommend any specific tactic, potential methods might include hacking Internet and mobile phone based communications between LeT and periodicals/news media.

---

[9] This complex option (and others) are discussed in greater detail in Chap. 11.

Observation 4. Encouraging resignations of LeT leaders and maintaining a Pakistani government ban on LeT has been flagged as a key factor in several policies. Such resignations appear to go hand-in-hand either with internal leadership conflicts within LeT or when one element of the LeT was officially banned, leading to certain leaders announcing that they are stepping down. Likewise, bans place restrictions on LeT (though often cosmetic, they are still restrictions) that appear to dial down LeT terror attacks of certain types. *Thus, this study recommends maintaining sustained international pressure to ban LeT and its various elements*—and even small steps that can hurt LeT's organizational effectiveness can be helpful.

Observation 5. Going hand in hand with the previous observation is that all four policies agree that the release of LeT prisoners by Pakistan must be stopped. This can only be done with *sustained international pressure.*

Observation 6. All four policies in Chap. 11 require that the operation of LeT's training camps be disrupted. LeT's training camps provide not only a sense of community to LeT recruits, but also a training ground for fedayeen attackers and those calling for violence. Taking covert action to disrupt these training camps places LeT on the defensive, diverting resources to continuing smooth functioning of their camps, instead of using those resources and time to plan terror attacks. Appendix H summarizes some of the LeT locations and training camps described in the open source literature.

Observation 7. All four policies studied in Chap. 11 suggest that the strategy of raiding LeT hide-outs, killing LeT personnel, arresting LeT personnel or issuing warrants for their arrest have had mixed results reducing LeT-backed terror attacks. This is borne out by the data presented in Chaps. 4, 5, 6, 7, 8 and 9 which show that while such counter-LeT operations been followed by a reduction in terror attacks on civilians and a reduction in armed clashes, they have also been followed by an increase in attacks on public sites and attacks on holidays. This is counter-intuitive and will be especially surprising to law enforcement. At the least, any such overt counter-LeT operations should be followed by a period of higher security and extreme caution, especially with increased protection of public sites and increased security deployments and intelligence operations on holidays.

Observation 8. Two of the four policies studied in Chap. 11 recommend that the relationship between LeT and other Islamist organizations (such as Jaish-e-Mohammed, al-Qaeda, and Hizb-ul-Mujahideen) be disrupted. The free flow of radicalized fighters between these organizations, and the cooperation between them, allows LeT to have access to resources that support their terror operations. These need to be disrupted. This option may be relatively easy to implement. Terrorist groups may cooperate, but also frequently feud and compete. LeT currently has an antagonistic relationship with al-Qaeda, has fought Hizb-ul-Mujahideen in the past, and frequently views Jaish-e-Mohammed as a rival. Encouraging further distrust, especially in view of the recent US Government announcement of a $10 M bounty on LeT leader Hafez Saeed, may well be worth considering (BBC News 2012).

Observation 9. Two of the four policies suggest disrupting the social services programs and medical programs run by LeT. These programs form a major backbone of citizen services in certain parts of Pakistan and help the LeT connect with the local population. The resulting population support improves LeT's ability to collect funds and recruits and must be disrupted in order to reduce LeT's access to resources. In Chap. 11, we recommend trying to generate alternative sources to provide social services and medical services to the Pakistani population through well-branded international NGOs. Additional funding could be made available to such NGOs though extreme care will be required to ensure the funds do not get inadvertently co-mingled with LeT's social service operations.

Observation 10. Often times, high profile overseas trials of LeT personnel (e.g., Faheem Lodhi in Australia) appear to inflame the passions of LeT supporters. This study recommends strongly that the publicity generated by such trials be kept to a minimum with few prosecutorial statements to the press and gag orders issued by the courts involved.

Chapter 11 advises that a combination of all these recommendations be used. Chapter 11 also makes specific *tactical suggestions* to achieve these goals which are summarized below.

- *Increased Non-Violent Covert Action.* Chapter 11 suggests non-violent covert action aimed at disrupting electricity (though Pakistan's electricity operators seem to be doing a good job at this already), telephone, Internet, and water supplies in and around LeT facilities, as well as covert cyber-action to increase mutual suspicion within LeT's ranks, potentially increasing the amount of intra-organizational conflict within LeT and fostering potential splintering within the organization. This, in turn, has the potential to reduce the resources and time currently available to LeT to plan terror attacks.
- *Increased Deterrance and Coercive Diplomacy.* Chapter 11 also suggests the use of coercive diplomacy as a means of holding Pakistan responsible for LeT terror attacks. Coercive diplomacy steps could include threats by India to "dial down" Pakistani water supply from the Indus river (the head waters of several of the major tributaries of the Indus river are controlled by India) if LeT carries out certain attacks, reduction in international aid to Pakistan if LeT carries out certain attacks, and trade sanctions. Public statements need to be made by anti-LeT entities in a way that they—and the Pakistani government—both know cannot be easily reversed if the conditions triggering the threat occur. This ensures a form of deterrence first suggested by (Schelling 1980).
- *Non-Violent Covert Campaigns against LeT's Communications Arm.* Chapter 11 suggests non-violent campaigns to disrupt LeT's ability to disseminate their message to their supporters. These include information operations, selective cyber-attacks, random jamming, and disruption of electricity and phone services around LeT's communications arm.

Chapter 11 discusses the pros and cons of each of the suggested policies in greater detail and also examines some of the tactics discussed in greater detail. However, it should be emphasized that the goal of this book is to understand LeT's behavior and suggest strategies to diminish terrorist acts carried out by them. The tactical suggestions made in this book are just suggestions—they require significant additional work and a more detailed understanding of LeT's network structure.

1.6 Conclusion

Generating good and non-violent behavior by LeT poses a major challenge that first requires understanding the circumstances under which LeT carries out terrorist actions, and then using that understanding to formulate carefully calibrated policies aimed at preventing those circumstances from arising.

This book presents a total of 61 Temporal Probabilistic (TP) rules that describe the conditions under which LeT carries out a variety of terrorist acts. These 61 TP-rules are a small sampling of over 15,000 TP-rules we derived.

A Policy Computation Algorithm was then used to automatically compute policies that can be used to engineer circumstances that make it difficult for LeT to carry out such acts. The LeT Violence Non-Eliminability Theorem in Chap. 10 shows theoretically that it is impossible to bring about circumstances that are likely to simultaneously eliminate all forms of terror carried out by LeT, but that most can be significantly reduced via eight policies, four of which are studied in Chap. 11—the other four are almost identical to the four that discussed. Each of these four policies is complex and perhaps this complexity explains why these policies have not been studied before in the literature, though policy makers have certainly known elements of these policies.

This study concludes with a note of caution. Should these policies be adopted and put into practice, it is likely that LeT will change its behavior. Models (and humans) are not omniscient, and predicting exactly what LeT will do, and how it will adapt, are major challenges. This study does not claim that the policies recommended in this book will put LeT out of business—only that it will make it operationally significantly harder for them to carry out terrorist acts for some time.

LeT's actions and the prevailing circumstances in their environment must be continuously monitored and the policies updated in real-time to keep up with LeT adaptations. Fortunately, the computational tools described briefly in Chaps. 3 and 10 make this possible with very little work on the part of counterterrorism organizations.

Last, but not least, it should be emphasized what this book does *not* do. It does not develop techniques or make predictions about individual attacks. Though this would be a desirable capability, this work is based on open source data and the complex intelligence information (usually classified) that might identify, predict, and disrupt individual attacks are not available to open-source researchers for rigorous analysis.

References

Abbas, H. (2005). *Pakistan's drift into extremism: Allah, the Army, and America's war on terror.* London: M.E Sharpe.

Baral, C., Kraus, S., Minker, J., & Subrahmanian, V. S. (1992). Combining knowledge bases consisting of first order theories. *Computational Intelligence, 8*(1), 45–71.

BBC News. (2012). US Puts $10M bounty on Lashkar-e-Taiba's Hafiz Saeed. April 3, 2012, http://www.bbc.co.uk/news/world-asia-india-17594018

Bell, C., Nerode, A., Ng, R., & Subrahmanian, V. S. (1994a). Implementing deductive databases by mixed integer programming. *ACM Transactions on Database Systems, 21*(2), 238–269..

Bell, C., Nerode, A., Ng, R., & Subrahmanian, V. S. (1994b). Mixed Integer methods for computing non-monotonic deductive databases. *Journal of the ACM, 41*(6), 1178–1215.

Brenner, J. (2011). *America the vulnerable.* New York: The Penguin Press.

Craig, G., & George, A. (1990). *Force and statecraft: Diplomatic problems of our time* (2nd ed.). New York: Oxford University Press.

Dickerson, J., Mannes, A., & Subrahmanian, V. S. (2011). Dealing with Lashkar-e-Taiba: A multi-player game-theoretic perspective. *Proceedings of the IEEE International Symposium on Open-Source Intelligence and Web Mining*, Athens, Greece, September 2011.

Fair, C. C. (2009). *Antecedents and implications of the November 2008 Lashkar-e-Taiba (LeT) attack upon several targets in the Indian mega-city of Mumbai.* Santa Monica: Rand Corporation. March 11, 2009 http://www.rand.org/pubs/testimonies/CT320/

Government of India (2008) Mumbai terrorist attacks: Dossier of evidence. *The Hindu*: Online Edition of India's National Newspaper. http://www.hindu.com/nic/dossier.htm

Gupta, S. (2011). *The Indian Mujahideen: Tracking the enemy within.* India: Hachette.

John, W. (2011). *Caliphate's soldiers: The Lashkar-e-Tayyeba's long war.* New Delhi: Amaryllis and the Observer Research Foundation.

Krepon, M. (2010). The Perils of proliferation in South Asia. *Arms Control Today*, April 2010 http://www.armscontrol.org/act/2010_04/BookReview

Minker, J. (1982). On indefinite databases and the closed-world assumption. *Proceedings of the 6th International Conference on Automated Deduction, Lecture Notes in Computer Science*, Vol. 138, pp. 292–308.

Rana, M. A. (2006). *A to Z of Jihadi Organizations in Pakistan* (Saba Ansan, Trans.). Pakistan: Mashal Books. http://www.desistore.com/jehadiorg.html

Rotella, S. (2011). *Pakistan's terror connections: Chicago terrorism trial what we learned and what we didn't, about Pakistan's terror connections.* ProPublica. http://www.propublica.org/topic/mumbai-terror-attacks/

Rubin, A. (2010). Militant group expands attacks in Afghanistan. *The New York Times*, June 15, 2010. http://www.nytimes.com/2010/06/16/world/asia/16lashkar.html?pagewanted=all

Schelling, T. C. (1980). *The strategy of conflict.* Cambridge: Harvard University of Press.

Subrahmanian, V. S. (2007). Cultural reasoning in real-time. *Science, 317*(5844), 1509–1510.

Subrahmanian, V. S., Albanese, M., Martinez, V. M., Reforgiato, D., Simari, G., Sliva, A., et al. (2007). CARA: A cultural adversarial reasoning architecture. *IEEE Intelligent Systems, 22*(2), 12–16.

Tankel, S. (2011). *The threat to the U.S. Homeland Emanating from Pakistan.* Washington, DC: Carnegie Endowment for International Peace, May 3, 2011, http://carnegieendowment.org/files/0503_testimony_tankel.pdf

Tankel, S. (2011b). *Storming the world stage: The story of Lashkar-e-Taiba.* London: C. Hurst & Co.

Tellis, A. (2010). *Testimony by Ashley J. Tellis. Bad Campany-Lashkar-e-Tayyiba and the Growing Ambition of Islamist Militancy in Pakistan.* Washington: United States House of Representatives, Committee on Foreign Affairs, Subcommittee on Middle East and South Asia, March 11, 2010, http://carnegieendowment.org/files/0311_testimony_tellis.pdf

The Age. (2006). *Lodhi 'Deserves' 20 Years.* August 23, 2006 www.theage.com.au

United States Government, National Counterterrorism Center, Worldwide Incidents Tracking System. (2011). *Criteria*, http://www.nctc.gov/witsbanner/wits_subpage_criteria.htm

Chapter 2
A Brief History of LeT

Abstract This chapter provides an overview of LeT from their creation to the end of 2011. It describes the goals of the group, other groups in their ecosystem, the types of attacks they have carried out, the internal dynamics of the group, and the relations they have with the Pakistani military and civilian government. It also includes brief profiles of selected LeT leaders.

This chapter provides an in-depth overview of the organization, history, and operations of Lashkar-e-Taiba (LeT). The first section describes of the group's foundation and the political and social context of LeT's founding, along with a discussion of LeT's ideology and worldview, and a profile of its founder and leader Hafez Mohammed Saeed. The second section describes LeT operations in Pakistan, including its organizational infrastructure which includes a social welfare arm that administers a network of schools and medical facilities, a communications arm that holds rallies and publishes several magazines, and an extensive fundraising operation. The section also discusses LeT's recruitment strategies and its relationship with the Pakistani government. The third section describes LeT operations against India in Jammu and Kashmir.[1] This section begins with a description of Kashmir's history and geography, and is followed by a survey of how LeT tactics have evolved in Kashmir from massacres, to fedayeen strikes, to hit and run attacks. This section concludes with a description of the infrastructure needed to support LeT's armed operations. The fourth section studies LeT operations in the rest of the Indian sub-continent, beginning with LeT's first major strike into India itself against the Red Fort in Delhi in December 2000. The section discusses the growing militancy among India's Muslim population, and LeT's relationship with Islamist groups in India. The section describes major LeT backed operations in India such as the 2006 Mumbai train bombings (attributed by

---

[1] Throughout this book, we use the term "Jammu and Kashmir" to refer to that part of Kashmir administered by India. Likewise, Pakistan-administered Kashmir (or Azad Kashmir) is used to refer to that part of Kashmir currently controlled by Pakistan.

V. S. Subrahmanian et al., *Computational Analysis of Terrorist Groups: Lashkar-e-Taiba*, DOI: 10.1007/978-1-4614-4769-6_2, © Springer Science+Business Media New York 2013

different authors to LeT directly or to the Indian Mujahideen which is closely linked with LeT) and the 2008 siege of Mumbai. This section concludes by examining LeT's growing operations in Afghanistan. The fifth section is an overview of LeT's international operations including its links to Islamist terrorists around the world, including al-Qaeda, and the activities of LeT operatives worldwide such as Willie Brigitte and David Coleman Headley. This chapter concludes with a brief discussion of the major questions about LeT's future intentions and strategies.

2.1 Origins and Overview

2.1.1 What's in a Name?

Lashkar-e-Taiba (which is variously translated from Urdu as Army of the Pure, Army of the Righteous or Army of the Good) is generally abbreviated as LeT. However it has been known by several other names as well. This is typical for terrorist organizations. Organization name changes and operating through front groups creates ambiguity that can confuse investigators and allow the terrorist groups and their sponsors to have plausible deniability about terrorist attacks.[2]

Initially the group was established in 1986 when Zaki-ur-Rehman Lakhvi merged his militant group with Jamaat ud-Dawa (a small Ahl Hadith missionary group founded by Hafez Saeed and Zafar Iqbal) to form the Markaz al-Dawa Irshad (Center for Preaching and Guidance—MDI) (Tankel 2011a). MDI established an armed wing, Lashkar-e-Taiba a few years later. Though the exact date of LeT's establishment is not completely clear, most analytical histories place its birth around late 1989 or early 1990.

In December 2001, after LeT and Jaish-e-Mohammed were accused by India of carrying out an attack on India's parliament,[3] the United States designated both groups as "Foreign Terrorist Organizations" (FTO) (U.S. Dept. of State August

---

[2] A short article describes the emergence of the Pakistani terrorist group Jaish-e-Mohammed (JeM) from other groups and how these groups changed their names after one version of the group achieved sufficient notoriety to be classified as a terrorist group by the United States— http://www.satp.org/satporgtp/countries/india/states/jandk/terrorist_outfits/jaish_e_mohammad_mujahideen_e_tanzeem.htm.

The shifting of names and use of front groups is common behavior for terrorist groups. Black September, the Palestinian terrorist group that carried out a number of high-profile attacks in the 1970s, most famously the attack on the Munich Olympics in 1972, was established by Fatah's intelligence unit but did not make public statements in order to avoid tarnishing Fatah's international image. (Karmon 2000).

[3] Although the Indian government held both LeT and JeM responsible for the Parliament attack, the operatives convicted in Indian courts for their role in this attack were all members of JeM. It is possible that individuals affiliated with LeT were involved, but concrete proof has not been advanced in the open source (Tankel 2011a, b).

16, 2011 and U.S. Dept. of Treasury October 13, 2011b). A month later, on January 13, 2002 under U.S. pressure, Pakistan's President Pervez Musharraf banned LeT. However, weeks before on December 24, 2001, LeT leader Hafez Saeed declared that LeT and MDI were now separated and that he no longer had any affiliation with LeT. Further, MDI reverted to the name Jamaat ud-Dawa (Society for Preaching—JuD). However, while LeT now supposedly restricted its operations to Kashmir, JuD offices in Pakistan were being used as LeT offices, suggesting the separation was in name only (Rana 2006). LeT has also worked through front groups such as Idara Khidmat-e-Khalq (Humanitarian Services Institution—IKK), which provides disaster relief both in Pakistan and also in the Maldives after the 2004 tsunami where it helped recruit LeT operatives (Roul 2010a, b). The United States Treasury declared IKK an FTO in April 2006 (U.S. Department of Treasury 2006).

Variations of this strategy repeated itself several times over the decade, with one important exception. In July 2004, JuD reportedly split over disputes within the group's leadership when top leaders were frustrated that Hafez Saeed was appointing family members to top posts in the organization, and established a new group Khairun Naas (which means "the most excellent of people" and also refers to the companions of the Prophet Muhammad—KN) (Rana 2004). Nonetheless, KN failed to emerge as a substantial group in its own right, with most of its leaders reportedly rejoining the LeT fold. Saeed still attempted to use this split to obfuscate LeT responsibility for the 2008 Mumbai attacks, citing the split and claiming that Zaki-ur-Rehman Lakhvi, the main alleged planner of the Mumbai attack, had been a member of the splinter group (Mir 2008a).

Other LeT splits followed the original pattern of being primarily cosmetic. In January 2009, a group calling itself Tehreek-e-Tahafuz Qibla Awal (Movement for the Safeguarding of the First Center of Prayer) held an anti-Israel protest in Lahore; yet the rally featured leaders of JuD and LeT and attendees waved the JuD flag (The Times of India 2009a, b). Similarly JuD established Tehreek-e-Tahafuz-e-Humat-e-Rasool (Movement for Defending the Honor of the Prophet) to organize protests against the Danish papers that published cartoons of the Prophet (The Economic Times 2010). More recently, JuD established a charity, Falah-i-Insaniyat to collect donations and provide aid to Pakistan's internal refugees from the fighting in the Swat Valley and later from the massive 2010 summer floods (Waraich 2009). Subsequently, LeT has spearheaded the establishment of Difa-e-Pakistan Council (Pakistan Defense Council), which has held mass rallies in Lahore, Rawalpindi, and Karachi protesting American and NATO activities in Afghanistan and Pakistan. JuD flags were waved by thousands of attendees and LeT's leader Hafez Saeed was a keynote speaker (Ahmad 2012) (Fig. 2.1).

The exemplar of this policy of artificial splits is that according to reports, in January 2009 the FBI determined that Lashkar-e-Taiba's Kashmir-based spokesman Abdullah Ghaznavi, who interacts with the media via anonymous phone numbers reportedly from Srinagar, was in fact the JuD's spokesman Abdullah Muntazir and based at the headquarters in Muridke (Mir 2009a, b).

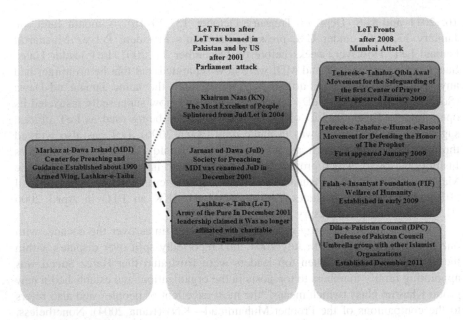

Fig. 2.1 Organizational affiliations

In the description that follows, unless otherwise specified, LeT and its fronts will simply be referred to as LeT.

2.1.2 Historical Context

The establishment of LeT occurred in the context of a number of historical and geopolitical factors within Pakistan, throughout the broader Muslim world, and internationally. At the same time, specific actions of individuals—leaders in the international Islamist movement, LeT's founders, and Pakistan's leaders also played a crucial role in LeT's establishment and growth as a major terrorist organization. In that regard, the 1979 Soviet invasion of Afghanistan and the alliances formed to support the Afghan resistance brought many of these factors together and created the conditions under which Lashkar-e-Taiba was founded and could flourish.

In December 1979 the Soviet Union invaded Afghanistan to stabilize the collapsing Afghan Communist government. Shortly thereafter, the United States decided to support the Afghan resistance in the hopes of making the Soviet occupation of Afghanistan as costly as possible. By necessity, neighboring Pakistan became a primary conduit for American support for the Afghan resistance (Coll 2004).[4]

Saudi Arabia, a key American ally, also supported the Afghan resistance. There were obvious geopolitical motivations behind Saudi support. Soviet control of Afghanistan (combined with the recent overthrow of the American-allied Shah of Iran) threatened to weaken America's position in the Middle East, leaving Saudi Arabia vulnerable to Soviet power.

Saudi support for the Afghans was primarily financial. Newly wealthy from the oil crises of the 1970s, the Saudis, as custodians of Islam's holy places of Mecca and Medina, were prepared to use their resources both to support their geopolitical ambitions and spread their version of Islam (The 9/11 Commission Report 2004). In Pakistan's President, General Zia ul-Haq, the Saudis found a willing partner. Zia had several motivations for Islamizing Pakistan. Zia had overthrown the elected Prime Minister Zulfikar Ali Bhutto and needed to establish his regime's legitimacy. Islamization was a national cause that would not be criticized and it shored up support from Pakistan's Islamist parties (Abbas 2005). Zia's use of Islam was not only instrumental, he was personally religious and believed that Islamization would strengthen the country morally and help it better deal with its many economic and social problems (Government of Pakistan nd). Many of Zia's reforms, which called for instituting strict Islamic penalties such as stoning for adultery, were never implemented. However, sectarianism and oppression of religious minorities increased and the government began officially supporting madrassas (religious schools) financially and granting formal standing to the degrees they issued. Substantial funding for these madrassas came from Saudi Arabia and other Gulf states (Abbas 2005).

Zia's Islamization campaign occurred in the context of a broader resurgence of Islam throughout the Muslim world. For much of the twentieth century, political leadership in most of the Muslim-majority countries of the Greater Middle East was predominantly secular in orientation. By the 1970s many of these regimes were seen as failures—in both the Arab-Israeli conflict (Sageman 2004) and in providing economic prosperity and basic freedoms to their people. Other factors were important as well. The Islamic revolution in Iran inspired many Islamists to consider the possibility of political revolution (The 9/11 Commission Report 2004). At the same time, the rise of an Islamist Shia power was a cause of concern to the Sunni regime in Saudi Arabia, which responded by exporting its interpretation of Islam throughout the world. In Pakistan, the Iranian revolution inspired Pakistan's large Shia minority to be more assertive in the public sphere, leading to a reaction by Pakistan's Sunni majority that was exacerbated by the government's Islamization policies (Abbas 2005).

---

[4] The story of the Soviet occupation of Afghanistan and the alliance between the United States, Pakistan, and the Afghan resistance has been told in many places, including from a Pakistani perspective (Youssef and Adkin 1992).

   This combination of events created a fertile ground for the rise of new Islamist organizations such as LeT. The official Islamization campaign in Pakistan, supported by Gulf Arab money, drew Pakistanis to Islamist organizations. The Afghanistan campaign's importance to Muslims worldwide drew Islamic activists and organizations to neighboring Pakistan, the staging ground for supporting the Afghan resistance. These activists inspired and aided Pakistani Islamists.

   The war in Afghanistan also had an important effect on Pakistan's leaders. Indian journalist Praveen Swami explains that supporting the American covert proxy war against the Soviets taught Pakistan's leadership that this strategy could be used to weaken an opponent, while "being calibrated to a point where it was not worth the while of the adversary to punish the sponsor-state by going to war." Pakistan sought to employ this means against its long-standing rival, India, particularly (but not exclusively) in Kashmir, and began nurturing the necessary proxies (Swami 2007). Another lesson the Pakistani military took from the war in Afghanistan was that hardline Islamists were more effective fighters then moderates (Yousef and Adkin 1992), so that state support was directed to Islamist groups such as LeT.

2.1.3 Abdullah Azzam, the Islamist Internationale, and the Founding of LeT

Abdullah Azzam rode this wave of history. Born near Jenin in the West Bank in 1941, Azzam received a doctorate in Islamic jurisprudence at Egypt's al-Azhar University in 1973 where he met Sheikh Omar Abdul Rahman, the future leader of Egypt's al-Gamaa Islamiya. Rahman is currently imprisoned in the United States for his role in plotting bombing attacks in New York in the 1990s (Bergen 2001).

   Azzam felt that the Muslims of the world needed to return to armed jihad, for example exhorting an audience during a tour of the United States in 1988:

   "Whenever jihad is mentioned in the Holy Book, it means the obligation to fight. It does not mean to fight with the pen..." (IPT Investigative Project on Terrorism 2008).

   Teaching at a University in Saudi Arabia, Azzam inspired many students including Osama bin Laden. In 1980 Azzam met some of the Afghan mujaheddin fighting the Soviets and decided to devote his energies to supporting the Afghan jihad. He moved to Pakistan where he was initially a lecturer at the Islamic University in Islamabad, before settling in Peshawar. Azzam travelled the world to raise money for the Afghan mujaheddin, recruit Muslims worldwide to fight in Afghanistan, and facilitated their travel (Bergen 2001).

   Based in Peshawar, near the Pakistan-Afghanistan border, Azzam established Makhtab al-Khidamat (Services Office) to coordinate the transfer of recruits to fight in Afghanistan. One of his top organizers and fund-raisers was Osama bin Laden (Bergen 2001). By the end of the decade, Azzam's Services Office had dozens of bureaus throughout the world (including Europe and the United States) that recruited volunteers, facilitated their travel, raised money to support Islamist causes, and spread Islamist propaganda (Emerson 2002).

While living in Pakistan, Azzam met Hafez Saeed, a professor of Islamic studies at the University of Engineering and Technology in Lahore. Around 1985–1986 Azzam, Saeed, and Zafar Iqbal (also a professor of Islamic studies at the University of Engineering and Technology in Lahore) merged their nascent Ahl Hadith missionary group JuD with Zaki-ur-Rehman Lakhvi's militant group to establish the Markaz al-Dawa wal Irshad (MDI), which translates as Center for Religious Learning and Propagation, in order to "organize the Pakistanis participating in Afghan Jihad on one platform" (Kohlmann 2000). Saeed, already a distinguished religious scholar, was essential to bringing religious legitimacy to the new organization (Tankel 2011b). There is an unproven rumor that bin Laden provided seed money to establish LeT (John 2011). One of the organization's first projects was establishing a large center at Muridke, a commercial town near Lahore.

MDI established its armed wing, Lashkar-e-Taiba (Army of the Pure) a few years after its founding. According to (John 2011), LeT was established on Feb 22, 1990. Although it established Afghan training camps in Kunar and Paktia, its participation in the Afghan jihad was limited. Only five LeT operatives were killed fighting in Afghanistan and LeT began to withdraw from Afghanistan due to fighting amongst the Afghan mujaheddin (Rana 2006). LeT focused its efforts on another front, the ongoing dispute between Pakistan and India over Kashmir.

2.1.4 LeT's Top Leaders

People who have met Lashkar-e-Taiba's founder and leader, Hafez Mohammed Saeed often describe him as jovial and academic in appearance (Kashmir Herald 2002). He is reportedly a gracious host, pressing food on his guests (Fisk 2010). Women who have met with him report that they were required to cover themselves so that only part of their face was visible and there was no risk that a strand of hair would be exposed (Stern 2003).

In the public arena, Saeed is a dynamic speaker who regularly addresses audiences of thousands. Mosques where he appears frequently overflow, with large crowds standing outside listening from loudspeakers. Videos of his sermons show throngs converging to hear him. Saeed's voice rings with passion and elicits vocal responses from his audiences (We are Ahle Hadith 2011; ANImultimedia 2011; TV Gujarat 2010).

Hafez Muhammad Saeed's life has been shaped by the history of Pakistan. His family, which included a number of notable Islamic scholars, moved to Pakistan during the Partition of British India into the modern nation states of India and Pakistan. In the fighting accompanying the partition, 36 relatives of Saeed were killed. Saeed was born in Sargodha in Pakistan in 1950. Ultimately, the family resettled in the Mianwali district where a government land grant brought the family renewed prosperity. Saeed is one of seven children, five of whom are living today (Kashmir Herald 2002).

Born Mohammed Saeed, Hafez is a title granted to individuals who have memorized the Koran by the age of 12. By 1974 he had earned a pair of Masters degrees from the University of Punjab (Pakistani Leaders Online 2011). An accomplished scholar, Saeed was sent to Saudi Arabia for advanced studies where he earned a Masters degree from King Saud University. This brought him into contact with leading Saudi religious figures, contacts that would be significant later. Saeed also worked for the Pakistani government's Council of Islamic Ideology (CII) . This council advises the "legislature whether or not a certain law is repugnant to Islam, namely to the Qur'an and Sunna" (Government of Pakistan 2011). When Saeed worked for the CII in the 1980s, it was a primary instrument in Zia's Islamization campaign (International Crisis Group 2003a, b). For over two decades Saeed was a professor of Islamic Studies at the University of Engineering and Technology in Lahore.

Saeed is married to a maternal cousin and his father-in-law was a distinguished Islamic scholar. He has a son and a daughter. His son and son-in-law are both involved with LeT. Recently his son-in-law Khalid Waleed has been acting as a spokesperson on Saeed's behalf (Tehelka Magazine 2009). Reportedly, Saeed also married the widow of a Lashkar operative who died in Kashmir. Saeed stated that he married the widow, who is several decades his junior, in order to ensure that she was adequately supported (Stern 2003).

LeT co-founder Zafar Iqbal is not as well known as Saeed, an occasional source of tension within the organization. Like Saeed, Iqbal was a professor at the department of Islamic studies at the University of Engineering and Technology in Lahore. He was born in 1953 and has been LeT's finance director, playing a critical role in LeT fundraising, as well as directing LeT's extensive network of schools (U.S. Treasury 2011a). Iqbal views himself as Saeed's equal and has clashed with him over the group's leadership. In 1999, they clashed when Saeed appointed Hafez Abdul Rehman Makki (the son of one of Saeed's maternal uncles and the husband of one of Saeed's younger sisters) as External Affairs chief (Tankel 2011a, b). While this conflict was averted through negotiations and Makki became External Affairs chief, another clash occurred in 2004 when Iqbal accused Saeed of preferring members of his Gujjar caste. Iqbal was a member of the Arain caste. Iqbal and other critics of Saeed felt that he was consolidating LeT under his own family. Saeed's son Talha was married to Makki's daughter and Saeed's son-in-law Khalid Waleed, LeT's spokesperson, was believed to be working with automobile smugglers. In this conflict Iqbal also raised Saeed's much younger second wife, although apparently Iqbal had also taken a much younger second wife. This conflict led to LeT splitting, in which Iqbal established his own group Khairun Naas. Ultimately, Iqbal rejoined LeT, and there were allegations that the split was engineered by Pakistani intelligence to reduce LeT's profile (Rana 2004).

Makki's status within LeT appears to be continuing to increase. In April 2012 the United States government offered a $10 million reward for information leading

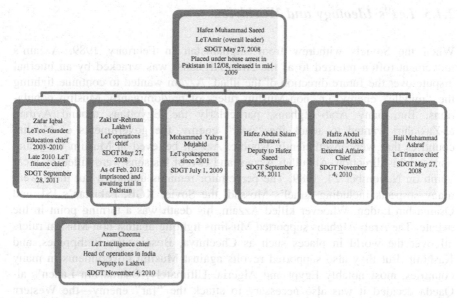

Fig. 2.2 LeT leaders & status

to the arrest of LeT's leader Hafez Muhammad Saeed. At the same time a \$2 million reward was offered for information leading to the arrest of Makki. Born in Bahawalpur around 1948, Makki has effectively become Saeed's top deputy and is believed to manage LeT's relations with al-Qaeda and the Taliban (Parashar 2012).

Zaki-ur-Rehman Lakhvi also played a role in the initial establishment of LeT. Born in 1960 in Okara, which is in Pakistani Punjab (U.S. Treasury 2008), Lakhvi was educated at Ahl Hadith institutions and was leading his own small Salafi jihadist group in Afghanistan in his early twenties where he came into contact with international Islamist figures (Roul 2012). Lakhvi and his jihadist group linked with the missionary group established by Saeed and Iqbal. Lakhvi, with no credentials as a religious scholar, did not become the public face of the organization (Tankel 2011). When the missionary group established an armed wing in 1990 Lakhvi emerged as its head. He has been the operational commander, leading LeT's efforts in Kashmir and overseeing LeT's extensive training program. Lakhvi has multiple pseudonyms and is referred to affectionately as *chachaji* (uncle) by LeT operatives. Two of his four sons were killed fighting in Kashmir. Lakhvi has also reportedly been the mastermind of LeT's expansion into India, leading the efforts to build networks in India and helping to plan major attacks including the 2008 attacks in Mumbai. There are also allegations that he played a role in the 2006 Mumbai train bombings although LeT's role in that operation is unclear (Roul 2012).

India charged Lakhvi in the November 2008 assault on Mumbai and Pakistani authorities arrested him in December 2009. However, his trial has been delayed and Western intelligence sources believe he is continuing to direct LeT operations from prison by mobile phone (Laskar 2011) (Fig. 2.2).

*SDGT Specially Designated Global Terrorist by U.S. Treasury.

2.1.5 LeT's Ideology and Worldview

When the Soviets withdrew from Afghanistan in February 1989, Azzam's movement (often referred to as the Arab Afghans) was wracked by an internal dispute over the future direction of the jihad. Azzam wanted to continue fighting the "middle" enemy, i.e. non-Muslim rulers of predominantly Muslim populations. But many Arab-Afghans, particularly the Egyptians around Ayman al-Zawahiri wanted to attack the "near" enemy, the governments of Muslim countries that were insufficiently Islamic. Azzam believed in Muslim unity and opposed this direction. Abdullah Azzam and two of his sons were killed by a car bomb on November 24, 1989. The perpetrator remains unknown but there were many suspects including Israel's Mossad, the Soviet KGB, Pakistan's ISI, and Osama bin Laden. Whoever killed Azzam, his death was a turning point in the debate. The Arab-Afghans supported Muslims fighting against non-Muslim rulers all over the world in places such as Chechnya, Bosnia, the Philippines, and Kashmir. But they also supported revolts against Muslim governments in many countries, most notably Egypt and Algeria. Ultimately, Osama bin Laden's al-Qaeda decided it was also necessary to attack the "far" enemy—the Western powers that supported the allegedly corrupt insufficiently Islamic governments (Sageman 2004). LeT, however, has, for the most part, adhered to Azzam's worldview, focusing its efforts against India, a non-Muslim power that (in LeT's view) oppresses Muslims. It has refrained from attacking the Pakistani state and its links to sectarian violence in Pakistan have been limited. Publicly, LeT calls for Islamic unity and criticizes conflicts within the Muslim world as weakening the Ummah (international Muslim community) (The MEMRI Blog 2010a, b, c).[5]

LeT is an Ahl Hadith organization. Hadith refers to the teachings and deeds of the prophet, which—along with the Koran—serves as the guide not only for religious practice but also for law and society. The Ahl Hadith movement is closely linked with the Wahhabi or Salafi traditions. These traditions seek to follow the "pure" Islam practiced by the Prophet Mohammed and his initial followers. Ahl Hadith is not the dominant tradition among Pakistan's Sunni population (Sunnis account for about 75% of the whole, while a fifth of Pakistan's population is Shiite). The two dominant Sunni groups are the Deobandis and Barelvis. Both of these sects adhere to the Hanafi school of Islamic jurisprudence, although their interpretation varies. The Barelvi tradition embraces Sufi mysticism and local folk practices. Deobandism, which emerged in the late ninetenth century, was a reform movement, which sought to purge Islamic practice of these local customs. Ahl Hadith rejects the Hanafi school of Islamic jurisprudence for its interpretations that go beyond the initial teachings of Mohammed. The Ahl Hadith movement explicitly rejects Barelvi practices. They tolerate Deobandis, who reject many of the same Barelvi practices as Ahl Hadith. Although the first Ahl Hadith

---

[5] This statement was not unique; in his 1999 interview with Jessica Stern, Saeed stated that the Sunni-Shia conflict was not important.

groups arose in the Indian-subcontinent in the late ninetenth century, they became more prominent starting in the 1980s, because they (along with the Deobandi groups) received state support from the Zia regime. The Ahl Hadith groups in Pakistan have also received funding from Saudi donors (Sikand 2007).

LeT is one of 17 major Ahl Hadith organizations in Pakistan. Some of these groups participate in Pakistan's political system, while others argue that Pakistan's system is un-Islamic. Some groups believe that jihad is a military activity that must be undertaken by all Muslims; while other groups argue some can conduct jihad on behalf of the community. Other groups believe that jihad is primarily about reforming one's own behavior. LeT refuses to participate in Pakistani politics and heavily emphasizes jihad as a military activity that is compulsory for all Muslims. LeT has clashed with one of the largest of the other Ahl Hadith groups, Markazi Jamiat Ahl Hadith (MJAH) which participates in Pakistani politics and takes a less strident attitude towards jihad. In early 2001 Markazi Jamiat Ahl Hadith accused LeT members of attacking its members and at another point of attacking Barelvi girls. Before he established MDI, Hafez Saeed was invited to join MJAH (Rana 2006).

In LeT's worldview the doctrines of dawa (preaching) and jihad are both required. LeT has an extensive network for propagating its vision of Islam, providing religious training, publishing periodicals (Rana 2006), and publicly calling for the full implementation of Islamic religious law in Pakistan (The MEMRI Blog 2010a, b, c). The dawa mission is intimately linked with the jihad mission. LeT believes that jihad is necessary to purify Pakistan and bring about a proper Islamic state. LeT's vision of jihad is expansive. Although LeT's early violent activity focused on the Kashmir conflict, this was never the organization's ultimate objective. In 1997 LeT leader Hafez Saeed stated, "We feel that Kashmir should be liberated at the earliest. Thereafter, Indian Muslims should be aroused to rise in revolt against the Indian Union so that India gets disintegrated" (Raman 1998). The August 2001 issue of the LeT periodical *Mujjala–ul-Dawa* included an article describing a jihad conference that stated, "Kashmir would become the doorway to jehad in Delhi, Agra, and Kathiawar." Not long before the 2008 Mumbai attack, Hafez Saeed addressed the Kashmir Solidarity Conference in Lahore and said it was an opportune time to take the war onto Indian soil (Mir 2008b). LeT's vision does not end with India. According to C. Christine Fair, who has reviewed LeT propaganda since 1995, the organization has long railed against the "Brahmanic-Talmudic-Crusader" alliance of Hindus, Jews, and Christians who were allied in an effort to destroy the Ummah (Fair 2009a, b). Email updates from LeT disseminated in 2000 and 2001 included frequent references to the Palestinians, a call for Pakistan to detonate a hydrogen bomb to "make the USA yield before Pakistan," and calls for jihad everywhere Muslims are oppressed including Chechnya, Kosovo, and the Philippines (Kohlmann 2006). Unsurprisingly, Israel and Jews are frequent targets of LeT's rhetoric. During Israel's 2009 war in Gaza, LeT organized protests against Israel (The Times of India 2009a, b). More recently, LeT protested a planned mass beard-shaving program in Karachi organized by Gillette to set a Guinness World Record. LeT referred to the program organizer as the "Jewish Company Gillete" (The MEMRI Blog 2010a, b, c).

According to Jessica Stern, when she interviewed Hafez Saeed in 1999, he presented his international economic agenda in social justice terms claiming that the West uses the World Bank and the IMF to oppress Muslim countries and that globalization is "a prelude to occupation." In that meeting another LeT leader complained that "Jews are brutalizing Muslims all over the world (Stern 2003)."

When questioned about Islamic terrorist attacks on civilians, the standard LeT response is that it was a plot by the CIA or Mossad. An LeT leader told Jessica Stern that "Anyone who goes against America is labeled a terrorist." He went on to state that the 1993 World Trade Center bombing must have been a CIA operation. After 9/11 LeT issued the following statement:

> ...Prof. Hafiz Muhammad Saeed has categorically stated that the terrorist attacks in the American cities were the doing of the Zionists and no Jihadic organization could be involved in such an un-Islamic activity. He said that America was planning [a] massacre of the Afghans under the cover of bin Ladens alleged involvement in the latest terrorist activity. He however added that such an action by Americans would invite even greater divine wrath... Hafiz Muhammad Saeed said that the Zionists and Christians were trying their best to link Jihad with terrorism through their powerful propaganda machinery. He said that undue American interference in the affairs of other countries was bound to have its repercussions (Kohlmann 2006).

More recently, LeT has responded to the Mumbai attacks with a similar trope. A spokesman for Saeed stated, "The Mumbai bloodbath seems to have been carried out by those who wanted to create more problems for Indian Muslims. Deccan Mujahideen is surely a fake name just to create confusion. We do know an Indian jehadi group with the name of Indian Mujahideen but this Deccan Mujahideen is a new thing, which is aimed at implicating the Muslims in the carnage. Even otherwise, let me make it clear that the followers of Hafiz Mohammad Saeed simply do not believe in killing innocent civilians" (Mir 2008a, b).

2.2 Pakistan: The Homebase

Depending on the political situation, the turn-off from the fabled and chaotic Grand Trunk Road (built by the emperor Sher Shah Suri who lived from 1486–1545) to LeT's headquarters ranges from poorly marked to completely un-marked at all. But, near the modest commercial town of Muridke, which is about 30 miles north of Pakistan's cultural capital Lahore, sits the Markaz (literally center or headquarters) of Jamaat-ud-Dawa, the charitable wing of LeT. Sprawling over about 200 acres, the compound includes (as visiting journalists regularly report) a profitable fish farm, cotton fields, horses, dormitories, schools, and medical facilities. The student body is reportedly over 1000 (some reports place it as high as 1600). The facilities include primary and secondary schools, as well as a college, which trains Alim (which is roughly equivalent to holding dual Masters in

Arabic and Islamic studies), all equipped with modern devices including computers. Evidence of martial activity varies with the political climate. Visits are carefully monitored and security guards armed with AK-47s patrol the grounds. In February 1999 Jessica Stern reported seeing an obstacle course intended to prepare students for Kashmir (Stern 2003) and others saw young men practicing martial arts. However, *The Dawn* reported in 2000 that people were educated for jihad at Muridke, but that actual training took place at camps near Muzzafarabad (Siddiqi 2000). With its well-manicured lawns and horses, the Markaz most resembles an elite private school—one reporter described it as "the Eton of Wahhabi Islam" and states that the complex was inspired by Aitchison College, one of Pakistan's elite private schools (Page 2008). The Muridke complex could also be described as a state within a state, LeT's attempt to create a perfect Ahl Hadith community. Un-Islamic behavior is forbidden, including smoking and music. Individuals who wish to live in Muridke show their commitment by smashing their television sets (Stern 2003). Establishing these enclosed communities is a common practice among Pakistan's Islamist movements. Jamaat-e-Islami has a housing society at its headquarters in Mansoora (Amir 2003) and Jaish-e-Mohammed was reportedly building a fortified complex at Bahawalpur (The Daily Times 2009), but LeT's center at Muridke is both larger and more secretive than any of the other religious communities.

The Muridke complex is only the largest individual component of LeT's vast organizational infrastructure throughout Pakistan. It is an infrastructure that includes offices, schools, medical services, and publications.

LeT has multiple departments concerned with outreach to different communities including teachers and students at colleges and universities, labor and farm organizations, a public relations department for managing press relations, and a foreign affairs department linking it with jehadi organizations worldwide. LeT runs its own schools including about 200 al-Dawa Model schools that provide primary education and eleven madrassas providing higher education. Emphasis is on religious studies, but science and modern subjects are also taught, albeit with an Islamist bent by teachers. About half of the schools are in Punjab (Pakistan's most populous province) with another 30 in the Sindh, and about a dozen each in Khyber-Pakhunkhwa and Azad Kashmir. Thirty-five of the al-Dawa Model schools in the Punjab are for girls, while all of the rest of the schools are for boys (Rana 2006).

Pakistani journalist Arif Jamal provides more detail. There are three types of schools run by LeT, none of which are (strictly speaking) madrassas because all of them include studies of other topics, particularly science. The al-Mahad Ala li-Dawah al-Islamiyah (High Institute for the Islamic Call) runs 16 institutions around the country, and provides higher education in Islam. There are also a network of schools that focus on memorizing the Koran, the Ma'az bin Jabl schools. This system was established in 1999 and now consists of five schools with over 100 students. The backbone of the LeT's education system are the al-Dawa Model schools which in 2002 had about 18,000 students and nearly 1000 teachers, including five schools for girls with about 5000 enrolled. All of the teachers at al-

Dawa schools are required to have had jihad training and many have fought in Kashmir. An important component to the growth of these schools has been the poor state of public education in Pakistan combined with the LeT's low fees compared to other private schools. There are schools in wealthy areas that (in 2002) might charge 1500 rupees per month (about $25) whereas schools in impoverished areas frequently charge about one-tenth of that and will waive fees for those who cannot pay (Jamal 2002).

The schools claim to teach the national curriculum and emphasize science, because it is, according to the principal of the school at Muridke, "Imperative that Muslims should learn science" (Bright 2008). However, the teaching has a heavy Islamist bent. Zafar Iqbal, then head of LeT's education department explained how LeT re-wrote the standard Elementary Reader to better indoctrinate students. He stated, "In the earlier Reader we had 'Alif' for Anar (pomegranate), 'Be' for Bakr (goat) and so on. This has been replaced by the concept of 'Alif' for Allah, 'Be' for Bandooq (gun), 'Te' for toop (cannon) and so on (Rana 2002)."

Although the LeT schools provide a large cadre of activists, only a relative few are dispatched to fight in Kashmir, or elsewhere. LeT ranks jihad and dawa as equally important and seeks to "convert" as many people as possible to its own school of Islam. Those who go through the LeT schools and are then sent on jihad training are believed to be more likely to return to their homes and proselytize on LeT's behalf (Fair 2009a, b).

LeT recruits from across Pakistani society. Many of those dispatched to fight in Kashmir are from impoverished backgrounds and have very little grasp of LeT's worldview, but are motivated by a combination of a desire for adventure (Swami 2005) (Abou Zahab 2007) or poverty. The lone survivor of the Mumbai attackers, Ajmal Kasab, joined LeT because he sought to learn weapons skills and embark on a criminal career and only later sought the fame of being a jihadi (The Economic Times 2008). When young men decide to sign up for jihad, it is relatively easy to find one of LeT's hundreds of offices throughout the country (Rana 2006).[6]

LeT does not recruit heavily from Pakistan's madrassas because they are primarily Deobandi and the Ahl Hadith madrassas are controlled by MJAH, which does not share LeT's position on jihad. LeT does recruit individuals with secular educations, deeming them better motivated and more capable (Jane's Islamic Affairs Analyst 2009). LeT also targets individuals from the wealthier upper classes for recruitment. However such recruits are valued for their skills (often they are engineers or are fluent in foreign languages) and are used to build the organization (Stern 2003).

---

[6] Rana reported that before January 13, 2002 (when JuD was banned in Pakistan for LeT's role in the attack on India's parliament) there were 1150 LeT/JuD offices in Pakistan. After the ban 116 offices continued to function. Other reports state that there are over 2000 LeT/JuD offices throughout Pakistan. See Rana, *A to Z of Jihadi Organizations*.

LeT also provides medical services, including a hospital at Muridke, mobile medical clinics, and an ambulance service (Al-Jazeera 2010). According to a 2003 report, the JuD's Al-Dawa Medical Mission has 2200 doctors that volunteer their services part-time, includes three hospitals and 47 dispensaries providing a wide range of medical services to thousands of Pakistanis who have little or no other access to medical care. The director of the program states explicitly that these operations both fulfill a religious obligation but also counter the activities of NGOs and Christian missionaries. Under Ahl Hadith rules, only women medical personnel can treat women patients so LeT has established a team of female doctors. LeT affiliated doctors are trained to proselytize as they provide treatment, hand out literature and maintain contact with patients after their treatment is complete. In addition, LeT digs wells to ameliorate water shortages (News 2003). These capabilities have allowed LeT to play a central role in disaster relief in Pakistan. With its training camps in Pakistani Kashmir, LeT (while suffering substantial casualties of its own) was one of the first organizations to provide aid after the 2005 Kashmir earthquake. LeT's rapid response, in comparison to the government's slow reaction helped increase LeT's public standing in Pakistan (Swami 2005). These efforts have been repeated in delivering aid to refugees from the Pakistani Army's 2008 offensive to re-take the Swat Valley from Pakistani Taliban, and in the wake of the devastating 2010 floods during which a JuD spokesman said the group had 2000 members working for flood relief, providing food, clothing and ambulances in NWFP and Punjab provinces. The volunteers were reportedly wearing badges for both JuD and its most recent incarnation Falah-e-Insaniyat (Shah 2010).

A central component of LeT's dawa mission is public outreach that includes print and online publications as well as public appearances by leaders at rallies, mosques, conferences and in the media. LeT publishes multiple publications appealing to different communities such as women and students. The publications feature calls for jihad against LeT enemies and articles that discuss the kinds of training LeT jihadis will receive. They also include general interest articles that attempt to make Islam relevant to modern life such as "Koran and Astronomy," and "Prophet's Medicine: Olive is the Cure for 70 Diseases (Bearak 2000a, b, c)." LeT has an active PR operation and its leaders and spokesman regularly communicate with Pakistan's print and electronic media including making appearances on major Pakistani television networks. On at least one occasion, in the aftermath of the summer 2010 floods, LeT's front group JuD purchased an advertisement in a Pakistani newspaper, calling for donations to support its flood relief efforts (Pakistan Media Watch 2010).

Overview of LeT publications (Rana 2008)

Publication	Circulation	Schedule	Language	Location & Audience
Mujalla-ul-Dawa	100,000	Monthly	Urdu	Lahore (banned by Pakistani government)
Ghazwa	20,000	Weekly	Urdu	Lahore
Al-Anfal	–	Monthly	Arabic	Lahore
Voice of Islam	–	Monthly	English	Lahore
Zarb-e Taiba	–	Monthly	Urdu	Lahore—youth & students (banned by Pakistani government)
Tayyibaat	–	Monthly	Urdu	Lahore—women
Babul Islam	–	Monthly	Sindhi	Karachi
Rozatul Atfal	–	Biweekly	Urdu	Lahore—children
Nanhay Mujahid	10,000	Monthly	Urdu	Lahore—children

LeT claims it sells 100,000 copies of Mujalla-ul-Dawa a month (sources at the press state that the regular run is actually 50,0000 to 65,000) (Sareen 2005)

LeT has been aggressive in employing modern technology to propagate its message. In the late 1990s LeT had websites that allowed individuals to make online donations and read articles and was disseminating its work through list-serves and email (Kohlmann 2006). It also ran an internet radio program al-Jehad (Sareen 2005) and more recently expanded onto Facebook. However, as LeT and its front groups attained higher public profile, their internet presence has been curtailed. Amir Mir reported that after the Mumbai attacks, Pakistani authorities removed LeT and JuD's English and Urdu websites, however other outreach operations continued unabated (Mir 2009a, b).

LeT also holds conferences and rallies throughout the country on a regular basis. The major annual rally, held at the headquarters in Muridke until 2002 (and later held in other locations in order to avoid too much official attention) attracts tens of thousands of attendees (Abou Zahab 2007). Reportedly, Osama bin Laden attended these annual conferences until 1993 and addressed them over the phone in the mid-1990s (Raman 2000). But this rally is only the LeT's largest. LeT clerics give sermons and hold rallies promoting LeT causes all over Pakistan. For example, in 2001, LeT sponsored a rally near the India-Pakistan border in which villagers were invited to hear the message of LeT veteran fighters and huge speakers broadcast their chants across the border into India (Rana 2006). In February 2010, after about a year of relative quiet by LeT leaders since the Mumbai attack, LeT organized conferences and its leaders spoke at mosques on Pakistan's Kashmir Solidarity day. Then, in June 2010 LeT held a series of rallies throughout the country against India's alleged theft of Pakistan's river water (Press Trust of India 2010a, b).

2.2.1 LeT's Finances

LeT's operations cost an enormous amount of money – consequently LeT has an extensive fundraising operation that that extends worldwide. According to one report in 2005, "Almost every third or fourth shop in all the major markets in Pakistan has donation boxes for jihadi groups (Sareen 2005)." Publication sales are a form of income, but the publications also contain exhortations to make contributions. Each unit of LeT runs fundraising operations in conjunction with the department of finance. In 2001 these campaigns raised approximately 200 million rupees (approximately $3.5 million.) Of course any information about LeT finances must be considered incomplete, but it does provide insight into the scale of the organization's operations. It is reported that many of LeT's wealthy donors, including wealthy Pakistani businessmen, do so anonymously in order to evade financial sanctions from the United States and international authorities. This particularly applies to donations from wealthy Gulf Arabs sympathetic to LeT's Ahl Hadith philosophy, which are believed to be a key source for LeT financing (Rana 2006).[7]

Donations to LeT (and other terrorist groups in Pakistan) are facilitated in a number of ways. After 9/11 international Islamist charities, such as the al-Haramain Foundation, al-Rashid Foundation, and the International Islamic Relief Organization have come under scrutiny for their role in supporting terrorist organizations worldwide, including LeT, (U.S. Dept. of Treasury October 5, 2011). Another method used is the hawala system, an informal financial transfer system that relies on trust. An individual in one city can hand cash to a hawala agent. That agent will contact his counterpart in another city, and on his word, the money deposited will be given to the recipient. There is no connection to the formal financial system (Ganguly 2001) (Jost and Sandhu 2000). The hawala system, while illegal, is a favored method of sending remittances to home countries around the world (Pakistani officials estimate that $7 billion flow through the country via hawala channels) annually and only a small portion is linked to terrorist activity (U.S. Dept. of Treasury (December 3 2010).

One fundraiser, commonly used by Pakistan's Islamist groups, and other charitable organizations is the collection of sacrificial animal skins on Eid al-Adha, Feast of the Sacrifice. On this festival, Muslims slaughter an animal and donate part of its meat to the needy and eat another part. In 2008, in Pakistan, about 5 million goats and sheep and 1 million cows were slaughtered. Organizations then collect hides and sell them to tanneries, providing about 45% of the needs of Pakistan's tanners. JuD reportedly collected over 100,000 skins and sold them for about $1.2 million (The Dawn 2010). In addition to corps of collectors, LeT uses a range of methods to convince people to participate in the sacrifice and donate the

---

[7] From 2001–2007 the Pakistani rupees was generally valued at about 60 to a US dollar so conversions are made at that rate. Public details about LeT financing are unlikely to be accurate, the purpose is to provide a glimpse into the scale and nature of LeT operations.

hide, including loudspeakers on top of mosques and an internet campaign that allows people to purchase a share of a sacrificed animal from anywhere in world. According to some sources, thousands of British Pakistanis contributed to this event, raising millions of rupees in 2003 (Sareen 2005).

Finally, the Pakistani government provides direct financial support to LeT. A September 2006 report in Pakistan's Herald magazine stated that large jihadi organizations such as LeT received monthly stipends of 2–3 million rupees ($30,000–50,000) (South Asia Terrorism Portal). Pakistan's civilian government has also provided financial support for LeT. In June 2010, when the provincial government of the Punjab published its annual budget, it had allocated 80 million rupees (about $1 million) to schools and hospitals affiliated with JuD, which was banned by the UN and the United States as a front for LeT. A spokesman for the Punjabi government insisted that it was overseeing the expenditure and that the operations were strictly humanitarian (BBC News 2010).

2.2.2 Recruitment and Training

Although LeT recruits from every region, ethnic group, and socio-economic class in Pakistan, the heart of LeT's support base is in northern and central Punjab, where there are even villages in which LeT is trusted to arbitrate disputes (International Crisis Group 2005).

LeT receives its greatest support from Punjabis and Mohajirs (individuals who are descendants of those who migrated from India to Pakistan during the partition of British India in 1947). LeT supporters are usually lower middle-class farmers and traders, LeT martyrs (operatives killed in fighting in Kashmir) write testaments of their life stories. These testaments are used as recruitment tools by LeT, in order to inspire others. They are also a useful source of sociological data about LeT. Mariam Abou Zahab surveyed 100 of these testaments from 2000–2003, taken from LeT's magazine *Mujjala–ul-Dawa* and also taped interviews of individuals recruited by LeT and planning to go to Kashmir as well as interviews with the relatives of the martyrs (Abou Zahab 2007). Many of the LeT recruits come from Punjabi families that migrated from the eastern, Indian part of the province, when partition occurred. This is also the background of LeT leader Hafez Muhammad Saeed. LeT also attempts to recruit from wealthier classes, but these recruits are usually not sent on jihad. Overall LeT recruits are better educated than those of other jihadi organization such as Jaish-e-Mohammed. Some LeT recruits have been to college, and only 10% had been to madrassas. The incentives to join LeT are substantial. Historically, the Kashmir conflict and tales of alleged Indian crimes against the Kashmiri people are a regular feature in Pakistan's media and education system, often with references to the horrors of partition. (With the Kashmir conflict moribund in recent years, fighting in Afghanistan is a now also a theme increasingly used by LeT for recruitment.) Joining LeT gave young Pakistani men an opportunity to take an active role in avenging these alleged atrocities and provided an opportunity to become famous. Finally, LeT operatives are well paid and the families of martyrs are rewarded. According to one report they are given a stipend of 1500 rupees—about $25 per month (Rana 2006).

LeT runs multiple training programs for its recruits. The daura-e amma and daura-e suffa programs are three-week intensive courses of religious education, open to any who wish to attend. Those who complete these programs and perform dawa and recruit their friends to LeT may be invited to the daura-e khassa, a three-month course of military training. Those selected to fight in Kashmir receive further, more intensive military training (Tankel 2011a, b).

The emphasis on religious training is further evidence that for LeT the dawa (proselytizing) function is just as important as jihad, and the two projects go hand in hand. LeT's recruitment process involves carefully cultivating the family of the recruit (particularly the mothers) and bringing the recruit's behavior fully into line with LeT's Ahl Hadith philosophy. Relatives report that recruits who underwent LeT training return with greater maturity and seriousness. LeT remains in contact with the families of martyrs and tries to keep them in the Ahl Hadith fold (Abou Zahab 2007).

2.2.3 LeT and the State

The scale of LeT's activities would be nearly impossible without at least the acquiescence of the Pakistani government and quite often the Pakistani government provides open support. Besides the financial support described above, Pakistani state support includes political and military aid.

The Pakistani military long viewed LeT as its favored proxy in Kashmir. According to Stephen Tankel, the Pakistani army and Pakistan's Inter Services Intelligence (ISI) agency trained LeT's trainers and designed LeT's training program (Tankel 2011a, b). In addition, from the 1990s and into the early 2000s the ISI, the army, and LeT worked together to plan attacks (Tankel 2011a, b). LeT primarily recruits from the same families and neighborhoods as the Pakistani military, leading to informal connections that facilitate this support (Riedel 2012). At times, infiltration efforts received covering fire from Pakistani forces. In other cases, the Punjabi government actively sought LeT's support. In 1998, several ministers both in the Punjabi government and the federal government visited Muridke and sought to enlist LeT's support against the sectarian violence that wracks Pakistan. LeT has had limited involvement in this violence (Abou Zahab 2007).

The Pakistani government and military have periodically clamped down on LeT, both in response to international pressure such as after the 2001 parliament attack and the 2008 Mumbai attack. The Pakistani government has also restricted LeT operations when it serves its own interests. According to Stephen Tankel, in 2004, when relations between India and Pakistan began to warm, LeT activities were restricted and the army began reducing the jihadi ranks. After the July 2005

London subway bombings in which two of the four bombers trained in Pakistan and one, Shezad Tanweer, may have had contact with LeT (The Guardian 2005), the Pakistani government began restricting the Kashmiri jihadists even more severely. Unauthorized infiltrators were arrested when they returned to Pakistan and in some cases their families were threatened if they infiltrated Kashmir (Tankel 2011a, b).

Nonetheless, the government of Pakistan always kept LeT (along with its other proxy forces) a viable option. Activities were restricted, but organizations were never fully closed down. The regular "house arrests" of LeT leader Hafez Muhammed Saeed illustrate the Pakistani military's double game. Although his movements were restricted and sometimes he was prevented from addressing large rallies, he remained active. For example, in September 2009 he attended an iftar (the feast with which Muslims break their fast during Ramadan) at the head-quarters of the 10th Corps of the Pakistani Army in Rawalpindi despite Punjabi police placing restrictions on his movements (The Times of India 2009a, b) (South Asian Terrorism Portal 2011). In the period since the Mumbai attacks, Pakistani courts have not authorized Saeed's continued detention due to insufficient evidence provided by the Pakistani government and made little progress in trying the seven LeT members it arrested in connection with the Mumbai assault (Curtis 2011).

Finally, whatever restrictions the Pakistani government has placed on LeT, unlike other Pakistani jihadi groups, LeT has never turned its guns on the Pakistani state.

2.3 Kashmir: The First Front

In the autumn of 1999, Indian commanders in Badami Bagh Cantonment, India's largest military base in Jammu & Kashmir, had some justified feelings of accomplishment. In July, the Pakistani army had been forced to withdraw from Indian bases they had seized in Kargil, a strategic mountainous zone along the Line of Control that divided Kashmir between India and Pakistan. Pakistan's rebuffed offensive had left that country internationally isolated and suffering from internal turmoil, while the state of Jammu and Kashmir was relatively quiet. However, in the evening of November 3, 1999 a team of LeT terrorists infiltrated this supposedly secure zone, entered the unguarded public relations office of the 15 Corps and killed the Public Relations Officer, Major Purushottam, and seven of his staffers. The LeT terrorists then held soldiers at bay for over 10 h before they were killed (Swami 1999) (Rediff.com 1999). LeT claimed to have killed 43 Indian soldiers, including a Colonel and two Majors, and that it had alerted the media using Major Purushottam's own phone (Rana 2006).

Although LeT had been sending fighters into Kashmir for the better part of the decade and its social activities within Pakistan were extensive, it had maintained a low profile in Jammu & Kashmir (Indian Express 1999). However, the attack on the 15 Corps HQ at Badami Bagh and other dramatic fedayeen[8] missions were seen as restoring Pakistan's wounded honor and helped propel LeT into the front ranks of jihadi organizations in Kashmir.

2.3.1 Kashmir in Time and Space

Kashmir has, since ancient times, been a strategic crossroads. Although the terrain is forbidding there are many key passes through the mountains and Kashmir sat astride important trade routes, linked to the great Silk Road, that have been traversed by caravans between east and west for thousands of years.

At the same time, it is a defensible area, which could be secured against all but the most intrepid invading armies. One noted sixteenth century chronicler, described Kashmir's advantageous strategic position:

> On all the sides mountains, which raise their heads to heaven, act as sentinels. Though there are six or seven roads, yet in all of them are places where if some old women rolled down stones, the bravest of the men could not pass (Kalhana's Rajatarangini 1900, translation by M.A. Stein).

With its commanding heights, possession of Kashmir provided an immense advantage to those seeking to infiltrate the Indian sub-continent. The British sought to secure Kashmir to prevent Russian infiltration during their rule, thus making Kashmir the focal point of the "Great Game" described in Rudyard Kipling's *Kim*.

Kashmir is also very beautiful. The Vale of Kashmir, an ancient lake basin about 84 miles long and 20–25 miles wide, ringed by towering peaks on all sides, is verdant and temperate. Under the Mughals and later the British, the valley became a favored refuge against the heat of the Indian plains. One Mughal emperor on his deathbed reportedly said he wanted, "Nothing but Kashmir." Sir Francis Younghusband, Britain's resident in Srinagar in the early twentieth century said, "Each spot in Kashmir one is inclined to think the most beautiful of all, perhaps because each in some particular exceeds the rest."

Kashmir is now a central point of contention between Pakistan and India. As Indian independence movements grew in the early part of the twentieth century, British India's Muslim leaders became concerned that in an independent Hindu-dominated India they would become a permanently disadvantaged minority. This led Muslim leaders to argue for two Indias, one consisting of the provinces with Hindu majorities and the other consisting of provinces with Muslim majorities. India's Muslim majority nation was referred to as Pakistan, an acronym for some of the Muslim majority provinces (Punjab, Afghania, Kashmir, and Sindh). In Farsi, Pakistan means "land of the pure".

---

[8] Fedayeen attacks are attacks in which the attackers expect to die, but unlike a suicide bombing, they do not die by their own hand but instead are killed in the fighting.

When the partition of British India occurred in 1947, both sides sought the accession of Kashmir, which was a semi-independent principality. Pakistan argued that as a predominantly Muslim province it should rightfully be part of the homeland for India's Muslim population. However, the maharaja of Kashmir was a Hindu and India's leader Jawaharlal Nehru was of Kashmiri descent and had forged a close personal relationship with Sheikh Abdullah, the most popular leader of the Kashmiri people.

The maharaja avoided making a decision. As these events were unfolding, Muslims in the Poonch district rebelled against the maharaja, a revolt that was put down sharply and with substantial bloodshed. In October 1947 Pashtun tribesman from Pakistan's Northwest Frontier Province, hearing reports about the Muslim revolt and the ensuing harsh response, declared jihad and invaded Kashmir where they went on a rampage of looting (Schofield 1996). Praveen Swami, a leading Indian journalist on intelligence and security affairs argues that the 1947 invasion was a pre-cursor of future Pakistani efforts in which irregular forces were supported with advanced military equipment and intelligence while the government officially denied any role (Swami 2007). However, it is worth noting that in its early years, Pakistan's army was commanded by British generals, which limited the options of the leaders of the new Pakistani state (Haqqani 2005). Regardless, with the collapse of his rule the maharaja signed an agreement of accession to India and Indian troops arrived to stop the tribesmen.

The fighting left India in control of the greater part of Kashmir, including the Vale, which includes Srinagar (the largest city) and the majority of Kashmir's population. Pakistani forces controlled over 32,000 square miles. Mountainous Gilgit and Baltistan were incorporated into Pakistan as the Northern Areas, consisting of about 27,000 square miles. The Pakistani government organized the slender strip of over 5000 square miles it occupied directly across the 1948 ceasefire line (known as the Line of Control or LoC) as Azad (Free) Kashmir— Indians refer to this area as "Pakistan Occupied Kashmir" or POK.

Just as important as the physical outcome of the fighting is the perception of the events by the participants (each of whom regarded themselves as being in the right.) The core legal dispute is over the authority of the maharaja to sign the letter of accession to India. Pakistan argues that the maharaja did not possess this authority and that accession should have been determined by plebiscite. But the Kashmir question took on far greater resonance for both sides. From a strictly strategic standpoint, a Pakistani general observed, "Kashmir's accession to Pakistan was not simply a matter of desirability but of absolute necessity for our separate existence." He went on to explain that Indian control of Kashmir would allow Indian troops to easily cut critical transit routes and place India in control of crucial water sources (Khan 1970). Pakistan's founding leaders were despondent at the prospect of an Indian-controlled Kashmir. Pakistani leaders felt that India, aided by Britain, sought to stifle Pakistan's viability as an independent nation from the start (Schofield 1996). This perception, that Pakistan was never given a fair chance is a crucial component of the Pakistani worldview (Cohen 2004).

Further complicating this situation were the feelings of the Kashmiri people themselves. The maharaja's line had been established in 1846 when, under the Treaty of Amritsar a Hindu nobleman purchased the Vale of Kashmir from the British East India Company and merged it with the territories he already controlled—Jammu and Ladakh. Their century of rule had not been perceived as benevolent and the Muslim majority felt that they had been oppressed (Schofield 1996).[9]

After the partition, Pakistan's leaders began plotting a covert war in Kashmir designed to spark a popular revolt. Praveen Swami chronicles these efforts, but prior to the 1980s they had limited impact. In one important respect, Swami found Pakistan's strategy of subversion was effective. The low-level conflict helped imbue India's Kashmir policy with a pervasive sense of crisis that led to hardening positions and strong-arm policies (Swami 2007). It led to a policy in which, in the words of political science professor Sumit Ganguly, "the national political leadership, from Jawaharlal Nehru onwards adopted a singularly peculiar stand on the internal politics of Jammu and Kashmir: as long as the local political bosses avoided raising the secessionist bogey, the government in New Delhi overlooked the locals' political practices, corrupt or otherwise (Ganguly 1997)."

In addition, there were two large-scale open wars between India and Pakistan. In 1965 Pakistani forces attempted to seize Kashmir and were defeated. In 1971 India supported the independence movement in East Pakistan (now Bangladesh), which resulted in a clash between India and Pakistan. India was victorious and Pakistan was split into two countries. Although this war was not fought in Kashmir, it left Pakistan shattered and divided. Pakistani efforts to infiltrate Kashmir declined in the short-term, although the defeat also reinforced the feeling among many Pakistanis that they would never receive their due place in the international order. Pakistan's strategic establishment remained interested in undermining India. Events within Jammu and Kashmir created their opportunity.

With Pakistan's 1971 defeat, their allies in Jammu & Kashmir were discredited. However, a small cadre continued to support an independent Kashmir and carried out low-level attacks. In 1987 elections were held in Kashmir, but the results were disputed due to allegations of widespread fraud resulting in mass protests in Kashmir. Allegations of heavy-handed Indian responses, which allegedly included curfews, several incidents in which Indian soldiers allegedly fired on protesters killing dozens, and allegations of gang-rape by Indian soldiers, poured gasoline onto these fires. Kashmiris may have been further inspired in 1989 by the sudden collapse of the dictatorships of Eastern Europe (Habibullah 2008).

The leading pro-independence group, the Jammu & Kashmir Liberation Front (JKLF), expanded its operations and Jammu and Kashmir descended into chaos (International Crisis Group 2003a, b). Pakistan seized the opportunity to re-initiate

---

[9] For one account of the rule of the Hindu dogras (see Schofield 1996 pages 49–117) which includes multiple accounts of British officials calling for the rulers of Jammu and Kashmir to institute reforms and foster development.

operations in Jammu & Kashmir. Since its 1971 defeat, Pakistan had acquired substantial assets for a new round of conflict. First and foremost Pakistan was now a de facto nuclear power, which raised the costs of India applying its military superiority in a conventional war. Second, Pakistan had benefited greatly both in equipment, skills, and confidence from its alliance with the United States in the drive to push the Soviets out of Afghanistan. Finally, India was suffering from violent internal secessionist conflicts (most notably, rebellious Sikhs in Punjab, who were actively supported by Pakistan) (Swami 2007) (Gill 1997).

Pakistani intelligence worked with the JKLF, but found them unsatisfactory allies because their focus was on obtaining independence for Kashmir. Just as the Afghan conflict was winding down, Islamist groups in Pakistan began focusing on the Kashmir conflict. The Pakistani Directorate of Inter Services Intelligence (ISI) began cultivating Islamist groups as an alternative. The first major beneficiary was Hizb-ul-Mujahideen (HM) the armed wing of the Pakistani Islamist party Jamaat Islami. HM had a strong network of supporters in the valley and actually turned its guns on the JKLF (with Pakistani support.) As HM bore the brunt of Indian counter-insurgency, the ISI fostered other Pakistani Islamist groups including LeT (Swami 2007).

2.3.2 Initial Involvement and Massacres

According to LeT's propaganda, operations in Kashmir were initiated on January 25, 1990 with an ambush on a jeep carrying Indian Air Force personnel. Five pilots were killed. An LeT leader, stated that this target was chosen for symbolic reasons because it was the Indian Air Force that flew the Indian soldiers into Kashmir during the partition (Abbas 2002). As the JKLF reduced its militant actions and in 1994 laid down its arms, LeT expanded its operations. In 1993, LeT operatives struck an army base in Poonch and in 1994, LeT reportedly ambushed an Indian Army convoy, abducting and killing five. LeT operations were heavily focused on the Poonch and Rajouri regions of Jammu & Kashmir. This area is just across the Line of Control from Pakistani territory. It was relatively easy for LeT to infiltrate because the area is west of the Pir Panjal Mountains that protect the valley. The inhabitants of the area are ethnically similar to the Punjabis, the LeT operatives know the language and blend in more easily (Sikand 2007). This area also includes substantial Hindu populations, which LeT, in conjunction with other groups including HM, targeted with massacres. These massacres were intended to spark a Hindu exodus, fuel communal tensions, provoke reprisals, and intimidate local populations (Tankel 2011a, b). From the mid-1990s onward LeT carried out numerous massacres of Hindu communities. Praveen Swami, then the intelligence correspondent for the Indian daily *The Hindu* wrote, after a massacre in August 2000:

The massacres should surprise nobody, for the Lashkar-E-Taiba has traditionally used such violence to sabotage peace initiatives. Seven people were killed at BalJalaran in Rajouri on the eve of [then Indian Prime Minister Atal Behari] Vajpayee's visit to Lahore in February 1999, through the Wagah border, in the first of three communal massacres that night. Another family of four was killed at the Mohra Fata hamlet of Khorbani… a remote village in Rajouri district. Nine members of a family, three of them infants, were killed the same night at Barhyana in neighbouring Udhampur district. Signals intelligence personnel listening in on frequencies used by the Lashkar-e-Taiba heard controllers telling field units to "turn the snow red". The hamlets, which dot the hills of the Jammu province, are almost impossible to defend in strength.

In the wake of the Pokhran II nuclear tests [conducted by India] in 1998 [followed by Pakistan's nuclear tests], the Lashkar-e-Taiba used communal killings with good effect to signal the group's determination to use the new de facto parity between India and Pakistan to escalate violence. On July 27 that year, 17 villagers were lined up and shot dead at Sarwan and ThakrainHor villages in Kishtwar, Doda. Jewelry and cash were looted from the dead. The same group murdered 26 construction workers and injured 11 persons on August 3, 1998, at Kalaban…

Each communal massacre in the State has been met by Hindu communal mobilization. Critics have pointed out that both the Hindu and Islamic right feed off communal massacres. As mass killings provoke migrations, Hindus and Muslims tend to consolidate into ethnic ghettos, a development that obviously serves communal politicians well. Each backlash against the massacres, in turn, deepens the fissures between communities.... (Swami 2000a, b).

Despite LeT's bloodthirsty attacks, the overall situation in Jammu and Kashmir was beginning to stabilize. Civilian deaths, overall deaths, and number of incidents were declining and in 1996 elections were held in Jammu and Kashmir indicating the improving security situation (Habibullah 2008). India and Pakistan's civilian leaders had begun discussing confidence-building measures. At this point, in spring 1999, backed by several of its proxies including LeT, the Pakistani army seized Indian positions in the mountains overlooking the Kargil region. The Indian and Pakistani armies had an unwritten understanding that both sides abandoned their positions during the winter because weather conditions made the mountains uninhabitable. It was during this period that Pakistani troops, with support from the various Islamist proxies including LeT, took the Indian positions and fortified them (Habibullah 2008). The Pakistani incursion was meant to draw international attention to the Kashmir conflict. But, a vigorous Indian response and an absence of international support forced Pakistan to withdraw. Ultimately, the United States provided Pakistan with only a minimal, face-saving gesture (Abbas 2005).

2.3.3 Fedayeen

The first fedayeen attack was not the infiltration of Badami Bagh described above. It occurred several months earlier, in July 1999 shortly after the defeat in Kargil. A pair of LeT operatives attacked the Border Security Force Headquarters in Bandipura with grenades and gunfire—killing the Deputy Inspector General (LeT claims to have killed 13 Indian soldiers) (Rana 2006). Over the next several years there were dozens of these attacks including attacks on key security targets such as the Police Special Operations Group HQ in December 1999 in which 12 security personnel were killed, or a November 2002 attack on a Central Reserve Police Force Camp that killed six. However, LeT did not restrict its attacks to security targets. On January 5, 2001 a squad of six LeT operatives, using a stolen car from the State Forester, attempted to enter Srinagar Airport wearing police uniforms. Several months later in August, three LeT fedayeen began shooting inside the Jammu Railway station, killing 11; two of the fedayeen escaped (South Asia Terrorism Portal).

Overall, LeT has carried out many of these attacks in Jammu and Kashmir. These dramatic attacks spread terror while embarrassing Indian security forces (and restoring some of the lost Pakistani honor in the wake of Kargil and other defeats). They also burnished LeT's reputation as being in the vanguard of the jihad in Jammu and Kashmir. These attacks caused a controversy in Islamist circles because suicide is forbidden under Islam. However, LeT argued that fedayeen are not committing suicide, like those who strap bombs to themselves and detonate them. In fedayeen strikes, small units attack superior forces, typically with firearms and grenades and are prepared to fight to the death, but can and will escape if the opportunity presents itself. In their house organ *Mujala al-Dawa* in May 2001, LeT leader Hafiz Abdul Rehman Makki wrote that the fedayeen encounter death only at the hands of their attackers, not by their own hand (Rana 2006). In this, he argued that the fedayeen follow in the footsteps of the Companions of the Prophet whose actions are inherently legitimate.

Although LeT leaders claimed that they only attacked military targets, LeT in Jammu & Kashmir continued to engage in large-scale massacres. Perhaps the most notable of these massacres occurred on March 20, 2000, the eve of President Clinton's official visit to India. In that massacre LeT terrorists, aided by HM, massacred 35 Sikhs at Chattisinghpora, Anantnag. The attackers were dressed in Indian army uniforms. In an interview, one of the accused LeT killers stated, "*The Koran* teaches us not to kill innocents. If Lashkar told us to kill those people, then it was right to do it. I have no regrets." (Bearak 2000a, b, c).

LeT developed sophisticated bomb-making skills, planting IEDs targeting Indian security forces on roadsides, in abandoned vehicles, and overhanging branches. LeT also sought to disrupt the 2002 Jammu & Kashmir elections, issuing threats to discourage citizens from voting and assassinating political leaders (Chalk and Fair 2002). Expanding operations required local infrastructure and resources. Most of LeT's operatives were Pakistani and there was little support for LeT itself among the Kashmiris. LeT expanded its efforts to recruit Kashmiris, but with limited success. The Ahl Hadith practices were in opposition to the dominant Sufi practices in the valley and the sect was peripheral in Kashmir and had difficulty attracting adherents (Sikand 2007).

An article by N.S. Jamwal, a Border Security Force Commandant based on personal experience and interviews with colleagues, provides some insight into the logistical details of jihadi operations in Jammu & Kashmir (Jamwal 2003).

Following is a summary of the article's findings. Indian security forces have found Kashmiri terrorists have generally maintained operational security in their communications and use a mix of high and low tech means to communicate from satellite phones and hand-held radio sets with encryption technology, smoke signals and informal substitutions of words. Most of the funding for operations comes from the Pakistani wing of the organization or the ISI and is transferred to operatives via the informal hawala networks. Perhaps 10 % of the operating funds are raised in J&K itself, often by coercion and intimidation. Kashmiris serve jihadi networks as guides and by providing information (such as observing movements of local security forces), shelter, and food. Women are the preferred support operatives because they are less likely to be searched unless female security personnel are present. The threat of massacres or of threats to property are used to guarantee a given population will at least be neutral and not report terrorist movements to security forces. Terrorists have found shopkeepers in remote areas to be useful in providing information, supplies, and money. In some cases shopkeepers providing support for terrorists were identified when they were carrying goods such as expensive shoes, IED components, and dry fruits. Terrorist sympathizers who provide active support are known as "Over Ground Workers", or OGWs. The most useful OGWs are socially prominent figures who have extensive social contacts and can provide terrorists with useful information and connect them to resources. Kashmir, a vast area with varied geography, which includes forbidding mountains, dense jungles, and areas that are inaccessible for many parts of the year, creates multiple opportunities for terrorists to establish hidden bases and supply caches (Jamwal 2003).

2.3.4   After 9/11

In the wake of 9/11, international scrutiny focused heavily on Pakistan-based terrorist groups. This was augmented after the December 2001 attack on India's parliament for which the Indian government held JeM and LeT jointly responsible. In response, the Musharraf administration clamped down on cross-border infiltrations—although LeT had a sufficient presence in Jammu and Kashmir to continue operations without the infusion of new cadres and supplies from across the Line of Control. A positive January 2004 meeting between Pakistan's President Musharraf and Indian Prime Minister Atal Behari Vajpayee led to a resumed dialogue process and Pakistan sought to de-militarize some of its proxy forces. Violence never stopped, but decreased overall (although LeT's share of the violence in Jammu and Kashmir expanded). The July 2005 tube bombings in London, in which some of the bombers had links to Pakistan and may have had links to LeT again brought international pressure on Pakistan, which in turn reduced LeT activities. According to Tankel, at one point LeT operatives who infiltrated across the LoC without permission risked arrest and even their families were threatened (Tankel 2011a, b). LeT changed its tactics in Jammu and Kashmir to adapt to its reduced resources. According to Indian security officials, LeT was less likely to undertake complex attacks against hard targets in Srinagar. Instead, LeT was sending young men to toss grenades at soft targets (Fair 2007).

Although the intensity of the Kashmir conflict declined substantially, LeT remained active in J&K and occasionally carried out large-scale attacks. In winter 2009, only months after the 2008 Mumbai attacks brought LeT international notoriety, there were multiple gun battles between Indian security forces and LeT (The Daily Excelsior 2009). In January 2010, LeT carried out a fedayeen operation, storming a hotel at Lal Chowk in Srinagar leading to a 22-hour siege in which a policeman and a civilian were killed (ExpressIndia.com 2010). LeT continues to attempt to intimidate Kashmiris. In April 2012 LeT placed posters in the Pulwama district threatening local leaders as part of a campaign to undermine upcoming elections (Times of India 2012). Nonetheless, these dramatic incidents are the exception rather then the rule, by all metrics violence in Kashmir has declined dramatically since the height of the Pakistan-backed insurgency in the early 2000s. According to Indian government statistics in 2011 Jammu and Kashmir averaged less then one terrorist incident per day, as opposed to seven incidents on average per day in 2004. In 2011, 64 civilians and security were killed in Jammu and Kashmir, wheras in 2004 that number was nearly 1000. (Ministry of Home Affairs, Govt. of India 2012).

2.4 LeT in India

On December 22, 2000 LeT conducted its first fedayeen attack in India beyond Jammu & Kashmir when a pair of gunmen assaulted the Red Fort in Delhi leaving three Indians dead. The LeT gunmen entered the Red Fort complex after 9 PM, when the regular sound and light tourist show held at the Red Fort was finished, The attackers killed a civilian employee of the Indian military and a soldier, who served as the unit's barber. They then entered the Army barracks and killed another soldier, while injuring two others. After a 45-minute gun-battle, the LeT operatives escaped (The Hindu 2000). LeT quickly claimed responsibility for the attack, insisting the strike was on a military target since the Red Fort was administered by the military and there were military offices, interrogation centers, and barracks housing several hundred soldiers on the premises (Bearak 2000a, b, c).

The Indian government and people were shocked, but Indian security should not have been surprised. After the Kargil war, and a year before the Red Fort attack, Hafez Mohammed Saeed, had threatened to "unfurl the Islamic flag on the Red Fort (The Times of India 2003)." In fact, threats to attack the Red Fort had been a regular part of Islamist rhetoric since the partition itself. The Red Fort was not merely a military base or tourist attraction. It is a symbol of Indian identity, but it was also a symbol of the period of Muslim dominance in India. In the words of English author and journalist William Dalrymple, "The Red Fort is to Delhi what the Colosseum is to Rome or the Acropolis to Athens: it is the single most famous monument of the city. It represents the climax of more than six hundred years of experimentation in palace building by Indo-Islamic architects, and is by far the most substantial monument—and in its day was also by far the most magnificent—that the Mughals left behind them in Delhi." Dalrymple goes one to explain that during "the apex of Mughal power, the golden age when most of India, all of Pakistan, and great chunks of Afghanistan were ruled from the Red Fort in Delhi. It was an age of unparalleled prosperity: the empire was at peace and trade was flourishing (Dalrymple 1993)."

On the Eid after the attack, LeT re-enacted the Red Fort attack before a huge crowd at Gaddafi Stadium in Lahore (Abou Zahab 2007).

Pakistani intelligence and LeT had long sought means of encouraging disgruntled minorities within India to revolt, in the short-term to keep pressure on India and tie down its forces—in the long-term, according to Pakistani doctrine—to cause the state to dissolve. For example, in the late 1980s and early 1990s, Pakistan supported Sikh separatists in Punjab (Gill 1997). Pakistani intelligence and LeT believed that Indian Muslims were an ideal target for subversion because of their social and economic grievances against the Indian state.

Discontent among India's Muslims was stoked by the destruction of the historic Babri Masjid Mosque in Ayodhya in 1992 by a Hindu mob. This destruction was followed by communal riots in which over 2000 people (primarily Muslims) were killed. Again, the ISI sought to capitalize on India's internal disorder. The ISI worked with Mumbai crime lord Dawood Ibrahim (no longer believed to be in India, with unsubstantiated reports placing him in several Pakistani cities, Dubai, as well as Malaysia) who is believed to have engineered a series of 13 bombings across Mumbai that took over 250 lives on March 12, 1993 (King 2004). LeT was also beginning to make inroads and establish its own networks in India. In 1992, before the riots, Mohammad Azam Cheema, a former academic colleague of Hafez Muhammad Saeed was dispatched to India. LeT's first attack in India was held exactly one year after the destruction of the Babri Majid, a series of small bombings in several cities across India that killed two (Swami 2008a, b). While Indian police arrested some LeT operatives, LeT was not yet well-known known to Indian intelligence agencies (Tankel 2011a, b). However, as the decade continued, LeT built more extensive networks in India that carried out numerous low-level attacks, primarily bombings and stabbings. In a May 2000 article Praveen Swami describes the activities of an LeT cell in the city of Hyberabad that claimed to bomb theaters that showed pornography and attacked Hindu shopkeepers (Swami 2000a, b).

Essential to building networks in India were local recruits from India's Muslim population. One source for recruits was Indian Muslims who traveled to the Gulf states to work. In the Gulf they came into contact with Pakistanis and, if they expressed interest, could be surreptitiously transported to Pakistan for further training and indoctrination. LeT also established connections with organized crime syndicates run by Indian Muslims (such as Dawood Ibrahim) and built networks in Bangladesh and Nepal to ease communications with and move operatives into and out of India (Tankel 2011a, b).

LeT was also held responsible for the December 13, 2001 attack on India's parliament. In this attack, five terrorists entered the grounds of India's Parliament House in New Delhi. Parliament was not in session, but five police officers, a security guard and a civilian, along with the five attackers, were killed. India was outraged and held Jaish-e-Mohammed and Lashkar-e-Taiba responsible. The attack set off a lengthy confrontation between India and Pakistan in which hundreds of thousands of soldiers faced each other across the India-Pakistan border. Pakistan banned Lashkar-e-Taiba and Jaish-e-Mohammed, the first of many perfunctory bans. LeT formally denied responsibility for the parliament attack. LeT leader Hafez Mohammed Saeed claimed that LeT operations were more competently organized and that the Parliament was a civilian target and LeT does not attack civilians (Abou Zahab 2007).

While Hafez Saeed's denial may seem dubious, many who have studied LeT, including Stephen Tankel who did extensive field work interviewing LeT members, note that none of the individuals ultimately convicted for the Parliament attack were from LeT. Also, a former ISI head, speaking to Pakistan's Parliament, blamed JeM for the attack. However, Tankel explains it is possible that individuals with links to LeT provided logistical support for the attack (Tankel 2011a, b). This reflects the reality that the affiliations of India's Islamists were fluid. Other Pakistani Islamist groups were recruiting Indians, and many of the Indian Muslim recruits came from Indian Islamist organizations, leading to substantial interconnections between groups.

2.4.1 Local Allies: TIM, SIMI, IM and D-Company

Radical elements among India's Muslim population began organizing of their own accord. LeT's first Indian recruits came from Tanzim-Islahul-Muslimeen (TIM— Organization for the Improvement of Muslims) a self-defense group formed as communal tensions grew in the 1980s (Swami 2000a, b).

Ultimately, a far greater source of recruits was the Student Islamic Movement of India (SIMI), which was initially founded in 1977 as the student wing of the Islamic party Jama'at Islami Hind. However, inspired by the same events that were affecting Islamists everywhere, including the Soviet invasion of Afghanistan, the Iranian revolution, and Zia's Islamization policy in neighboring Pakistan, the SIMI activists became increasingly radical. These global causes were tempered by local concerns, particularly discrimination suffered by India's Muslims at the hands of the Hindu majority and a lack of access to jobs and education. Initially SIMI organized protests and distributed propaganda. But by the late 1980s, in the face of rising communal tensions, SIMI began moving towards violent confrontation with India's Hindu majority. After the 1992 destruction of the Babri Masjid Mosque and communal riots, SIMI began advocating for armed jihad in India. At its 1999 rally, one speaker declared, "Islam is our nation, not India (Swami 2008a, b)." However, the group was only banned after 9/11 because of its links to the Taliban and al-Qaeda (it challenged the ban in court, but without success). In its last public rally in 2001, it attracted 25,000 supporters and before the ban it had 400 workers (Sikand 2006).

LeT recruited from SIMI, developing a network to support its operations and providing training to SIMI members, although SIMI remained a distinct and independent organization (Tankel 2011a, b).

C. Christine Fair, investigating the current status of SIMI and a related group, the Indian Mujahideen (IM) found that many Indian security analysts believed that they were effectively the same group. Others analysts, including Fair, believe that the groups are linked, but distinct. In about 2004, IM emerged from the most radical members of SIMI and it engaged in an armed campaign. IM's members included a number of people with computer skills and many of its members had personal experience with communal violence. In 2007, IM began issuing manifestos via email and carrying out major attacks in its own name. Among these attacks were nine simultaneous blasts in markets in Jaipur in May 2008 that killed over 60 people. In July 2008, IM claimed credit for multiple synchronized bombings in Bangalore and Ahmadabad that claimed at least 40 lives. In Delhi in September 2008 five bombs in markets killed 30 people. Fair also notes that identifying the perpetrators of many attacks in India is difficult. Indian officials are sometimes accused of publicly crediting LeT as the culprit in order to emphasize the Pakistani hand, while de-emphasizing its domestic terrorism problem (Fair 2010). Stephen Tankel summarizes the LeT/IM relationship, writing, "...most Indian militants did not perceive themselves as proxies for either Lashkar or Pakistan.... In other words, the group [LeT] was a force multiplier for Indian militancy, rather than a key driver of it. Further, while Lashkar was the chief external outfit providing support for Indian jihadism, it was not the only one (Tankel 2011a, b)."

Another important possible source of support for LeT operations in India is believed to have come from D-Company, the criminal organization led by Dawood Ibrahim, the suspected perpetrator of the 1993 serial bombings in Mumbai. Although Ibrahim's whereabouts are unknown (though he is suspected to be in and out of Pakistan), his criminal network extends throughout south Asia and into the Middle East. Heavily involved with smuggling and particularly the heroin trade, Ibrahim is believed to be a major donor to LeT. Further, his smuggling networks have been used to help LeT move operatives and material into and out of India (Clarke 2010) (King 2004).

2.4.2 LeT Strikes India

While LeT built relations with Indian Islamists and helped them develop their offensive capabilities, it also targeted India directly to pursue its own agenda. On the afternoon of September 24, 2002 a pair of LeT gunmen stormed the Akshardham Temple, one of the largest temples in Gujarat. When LeT attacked, firing their guns and throwing grenades, the temple was packed with worshippers. Over thirty people were killed, including several security personnel, and more than seventy people were wounded before commandos killed the attackers the next morning about fourteen hours after they had entered the complex. Like other LeT fedayeen attacks, the two attackers came prepared for a long siege, carrying dozens of grenades as well as supplies of dried fruits and chocolate to maintain their energy levels (Joseph 2002). The attack occurred six months after communal rioting in Gujarat took hundreds of Muslim lives and was intended to avenge these attacks on India's Muslims (Press Trust of India 2010a, b).

LeT's next attack was also against a symbol of India's economic development. On December 28, 2005 a pair of LeT fedayeen entered an auditorium of the Indian Institute of Science campus in Bangalore during a conference. Throwing grenades and firing assault rifles, a professor visiting from Delhi was killed and five attendees were injured. The target was selected carefully as Bangalore had been at the center of India's booming high-tech sector and the Indian Institute of Science is one of India's foremost scientific institutions (Rajmohan 2006).

2.4.3 Targeting Mumbai

LeT's largest attacks however, were unleashed on Mumbai. A sprawling city, Mumbai is the financial capital of India and its expansion parallels India's growing international stature and economic power. Mumbai was no stranger to mega-terror. In March 1993, Pakistani intelligence working with the organized crime network known as D-Company (with close links to underworld don Dawood Ibrahim) set off 13 bombs across the city killing 257 people. Mumbai suffered many smaller terror attacks in the decades since including a double-car bombing (attributed by Indian security to LeT) in August 2003. One bomb was detonated near the tourist attraction Gateway of India and the other detonated near the jewelry market Zaveri Bazaar (Waldman 2003). In 2004 Indian security disrupted an attack planned on the Bombay Stock Exchange (in Mumbai) (Fair 2009a, b). On July 11, 2006 mega-terror returned to Mumbai when seven bombs exploded within minutes of each other during the afternoon rush hour in the first class sections of commuter trains throughout the city. Over 200 people were killed and over 700 were injured. The bombings were generally seen as a joint operation of SIMI and LeT, possibly with support from the ISI. The attack was both seen as avenging the anti-Muslim pogroms of 2002 and it occurred as Indian-Pakistani relations were warming. While Pakistan and India did go on to establish a formal counterterror information sharing mechanism, the achievement was more a symbolic an effort to keep the talks between the countries from being completely derailed (Tankel 2011a, b).

Although more people died in the July 2006 attack, the November 2008 siege of Mumbai captured international attention as television screens across the globe broadcast images of a major city under attack for several days. Previous LeT fedayeen attacks in India involved two man teams. The Mumbai assault included

10 gunmen who divided into five two-man teams. Departing Karachi by sea on November 22, 2008 the Mumbai attackers took their first Indian casualties when they hijacked the Indian fishing trawler MV Kuber and killed four of its crew. The captain was kept alive to pilot the ship closer to Mumbai. On the afternoon of November 26, when the MV Kuber arrived about four knots away from Mumbai, the attackers killed the fishing trawler's captain and boarded a small dinghy for their final approach. The attackers failed to sink the hijacked ship, which later provided evidence about how the Mumbai attack was organized (Government of India 2008).

The attack teams systematically divided up and hit a series of pre-planned targets including the train station, two of India's most iconic hotels, a restaurant known to be frequented by international visitors, and the Nariman House, a facility run by a Jewish religious group that provides meals and hospitality to Jewish travelers. The attack was carefully planned. Targets had been scouted beforehand. The focus on targeting Westerners and particularly Americans and Israelis was also notable. Most of the targets were well known as symbols of Mumbai as a cosmopolitan, international city. Targeting the little-known Nariman House highlights the extent to which LeT was willing to go to include Israelis among its victims. Additionally, the attackers used Thuraya satellite phones and Voice over IP technology to remain in contact with their Karachi-based handlers (Government of India 2008)[10] who monitored events on television and relayed intelligence to the gunmen in real-time. When Indian security forces regained control of the city after three days on November 29, over 166 innocent people were dead (along with 9 of the 10 attackers).

The targets and timing of the Mumbai assault highlight various LeT motivations. Striking major tourist hotels in Mumbai was an effort to undermine India's growing international prominence and wealth, both by targeting Indian elites and international travellers. In targeting the Nariman House, LeT was explicitly targeting the growing Indian-Israeli alliance. But, perhaps most significantly, the attacks occurred during a period in which some Pakistani leaders were attempting to pursue warmer relations with India. The Mumbai assault brought this process to an immediate halt (Riedel 2012).

Much of what is known about the planning of the Mumbai attack and the internal workings of LeT comes from the testimony of David Coleman Headley. Born to an American mother and Pakistani father, Headley had served prison time in the US for drug dealing, been an informant for the Drug Enforcement Agency, and ultimately fell into the orbit of LeT. A charming individual who was comfortable in Western society, Headley's American passport made him an ideal operative for a heavily monitored terrorist group, and conversely he embodied the nightmare of Western intelligence agencies. Headley scouted targets in Mumbai

---

[10] (Government of India 2008) Page 3 of the report describes the retrieval of GPS instruments and a Thuraya satellite phone. Page 12 of the same report describes the use of VoIP technology by the terrorists.

for LeT. Headley also travelled to Denmark to examine the feasibility of attacking the newspaper *Jyllands-Posten*, which, in 2005, published cartoons of the Prophet Mohammed—an act that outraged many Muslims worldwide. After his arrest, Headley pleaded guilty and agreed to testify against Tahawwur Rana, who ran a Chicago immigration-consulting firm that provided cover for Headley's travels. Headley's testimony, which was covered in extensive detail by *ProPublica* reporter Sebastian Rotella, revealed, among other things, that all of the group's major operators have ISI handlers (Rotella 2011a, b). Headley's testimony alleged the existence of LeT's "Karachi project," which, in the words of Indian analyst Animesh Roul, "entails Pakistan-based militant groups training and deploying Indian Muslims for attacks in the Indian heartland (Roul 2010a, b)."

Although the Mumbai attack led to LeT being banned in Pakistan and to international opprobrium, the organization continued to attack India. On February 13, 2010 a bomb went off near a popular German bakery in the city of Pune, killing 17 and injuring over 60, in an area frequented both by affluent Indians and foreigners. This blast may have been the work of LeT's local allies or it may indicate that LeT is in fact running a "Karachi Project" (Roul 2010a, b).

2.4.4 LeT in Afghanistan

Although LeT had been founded to support the Afghan jihad, when that campaign ended with the Soviet withdrawal, LeT also reduced its presence in Afghanistan. LeT avoided participating in the Afghan civil war and as an Ahl Hadith organization was not comfortable with the Deobandi philosophy of the Taliban. LeT's strongest allies in Afghanistan were the Salafis of the provinces of Kunar and Nuristan, who also had a difficult relationship with the Taliban (Tankel 2011a, b). LeT co-founder Zafar Iqbal told a reporter, "Taliban is a group of misguided people and we have a much higher standard and principles (Rana 2006)." After 9/11, as an organization, LeT attempted to steer clear of the fight in Afghanistan, but individual members sympathized with the Taliban and took leave to fight in Afghanistan. By 2004–2005, the organization began to provide formal support to its fighters heading to Afghanistan. One important source for this change was frustration in LeT's ranks at the organization's acquiescence to the American-Pakistani alliance. Allowing fighters to confront Westerners in Afghanistan was one outlet for this tension within the organization (Tankel 2011a, b). By 2008 LeT units were regarded as among the most effective confronting international forces in northeast Afghanistan. LeT forces participated in a July 2008 attack in which insurgent forces nearly over-ran a US base in Wanat, Nuristan (Jane's Terrorism and Security Monitor 2008).

More recently, LeT has turned to high-profile strikes against major Indian targets in Kabul, Afghanistan's capital. Here too, LeT is suspected of receiving ISI assistance. Pakistan casts a wary eye on India's growing presence in Afghanistan, concerned that a Delhi-Kabul alliance will leave Pakistan encircled by hostile states (Weinbaum and Harder 2008). These attacks included a July 2008 suicide car bombing of India's Embassy in Kabul in which 54 were killed including four Indian diplomats and guards. The attack was blamed on the Haqqani network, but it was LeT which had originally recruited the bomber (Swami 2008a, b). LeT is also suspected in a December 2009 bombing near a Kabul hotel hosting Indians. In February 2010 LeT operatives, along with gunmen from the Haqqani network, combined a carbomb attack and a fedayeen strike on guesthouses in Kabul hosting Indians. Eighteen people were killed including nine Indians. In an echo of the Mumbai assault, the attackers used cellphones to coordinate the attack with handlers outside of Afghanistan (Brulliard 2010) (Rubin 2010).

2.5 LeT International

In April 2004 the Indian newspaper *The Hindu* reported that British forces had arrested five LeT personnel near the Iraqi city of Basra. Their leader Danish Ahmed, was well known to Indian intelligence as an LeT recruiter and for coordinating LeT infiltration into India. LeT had rhetorically opposed the American intervention in Iraq, but the deployment of a high-profile operative such as Danish Ahmed was a sign that LeT was expanding its operations beyond the Indian subcontinent (Swami 2004). Although many LeT recruits sought to travel to Iraq, LeT did not ultimately send a large number of operatives to Iraq. Few LeT members were fluent in Arabic and LeT's links to the clandestine networks for smuggling operatives into Iraq were limited (Tankel 2011a, b). Nonetheless, the event highlights LeT's growing international presence. This international network has primarily been used to support activities on the Indian sub-continent, but it is a capability that could also be used should LeT choose to engage in international terror. Figure 2.3 presents a succinct summary of LeT's international operatives, while Fig. 2.4 shows a summary of some known LeT operatives in the USA.

2.5.1 LeT and the Islamist Internationale

LeT was integrated into the international Islamist terrorist movement from its foundation. Throughout the 1990s there was a complex network of Islamist organizations and individuals that sought to continue the jihad that started in the war against the Soviet Union. The network included formal organizations with extensive infrastructure such as LeT, clandestine organizations such as al-Qaeda, and Islamist charities that could direct recruits and resources to conflict areas while raising money and recruits from Muslim populations worldwide. The Arab Afghans felt that they had played a central role in the 1991 collapse of the Soviet Union and that they were riding the wave of history. Their call to reestablish the caliphate and the glory days of Muslim civilization through jihad had tremendous

Lashkar-e-Taiba International Chart

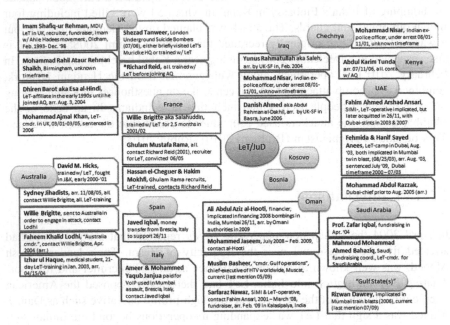

Fig. 2.3 LeT international

appeal to disaffected young Muslims worldwide, both in the greater Middle East, where many lived under oppressive corrupt regimes, but also in the West where some young Muslims felt disaffected from the modern secular democracies.

LeT was an important component of this network, both receiving assistance from around the world and providing support for other organizations. Peter Bergen describes meeting an LeT spokesman who had been examining a web site for the Chechen rebels, which LeT was supporting with its own fundraising efforts (Bergen 2001). Sheikh Abu Abdel Aziz (known as Barbaros for his thick reddish-orange beard), who was praised by Hafez Saeed for his efforts in helping to establish LeT by liaising with Saudi donors and establishing training camps, went on to fight in Bosnia and claimed to have fought in Africa and Kashmir as well (Post and Brand 1992). Foreign recruits came to LeT's training camp both to participate in the fighting in Kashmir, but also to learn skills they could take elsewhere. Charitable organizations such as the Saudi Arabia-based al-Haramain Islamic Foundation had offices all over the world and an estimated budget of $50 million. It provided financial and logistical support to al-Qaeda and other Islamist groups worldwide. Its office in Pakistan was designated as financing terrorist organizations including LeT (U.S. Dept. of Treasury, *Protecting Charitable Organizations*). Other components of this network were less organized, Azzam Publications was little more than an individual running a group of websites that

Lashkar-e-Taiba International Chart: USA

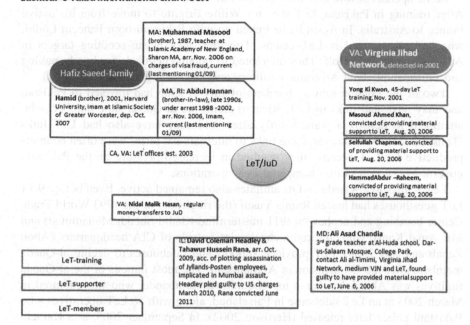

Fig. 2.4 LeT's US activities

directed donors and recruits to jihadi, organizations—one of his recommended groups was LeT (*Azzam Publications* 2001).

After 9/11, individuals seeking to join the Islamist jihad sought out LeT. Whereas al-Qaeda became less accessible, LeT retained a prominent public profile. LeT's new recruits included members of Pakistan's diaspora, other Muslims, and Western converts to Islam. LeT's status rose particularly with the Indian parliament attack which came so soon after 9/11 (although, as stated above, LeT was blamed, but is probably not responsible for this attack) and also because, with its recruiting offices throughout Pakistan, it was relatively easy for would-be jihadis from around the world to find them. As international counter-terror measures became more effective, jihadi hopefuls had greater difficulty joining LeT both because of international scrutiny, but also because LeT became rigorous in establishing the identity and intentions of recruits in order to prevent infiltration by Western intelligence agencies. Before 9/11 for example a group of Virginians could travel to Pakistani for training (they became known as the Paintball Jihadis because they maintained their skills playing paintball in suburban northern Virginia) (Associated Press 2004). Even after 9/11 Willie Brigitte, a French convert to Islam, could train with LeT in Pakistan before moving to Australia to establish an operational cell (BBC News 2007). However, later in the decade this became more difficult. An American from Georgia, Syed Haris Ahmed, traveled to Pakistan for training and was denied entrance into LeT camps (Associated Press 2009).

Some operatives with LeT links carried out, or nearly carried out terror attacks. After training in Pakistan, LeT paid for Willie Brigitte to move from his native France to Australia. In Australia he linked up with Pakistani-born Faheem Lodhi, who had also trained in LeT camps. Together, they began scouting targets in Australia on LeT's behalf. They attempted to obtain chemicals for bomb making and information about Australia's military and electrical grid (Brenner 2011).

Two of the London subway bombers of July 7, 2005 had traveled to Pakistan and may have had links to LeT. Richard Reid, the al-Qaeda shoe bomber, who attempted to destroy a plane shortly after 9/11 may have also had LeT links (Tankel 2011a, b). However, these links to international terror were often counterproductive for LeT because they resulted in increased pressure on the Pakistani government, which in turn hampered LeT operations.

LeT's links to al-Qaeda and its affiliates also remained active. Even before 9/11 LeT guesthouses had hosted Ramzi Yusuf (the ringleader of the 1993 World Trade Center bombing and nephew of 9/11 mastermind Khalid Sheikh Mohammed) and Mir Amal Kansi, who attacked a checkpoint outside of CIA headquarters (Abou Zahab and Roy 2004). LeT provided support and safehouses to fleeing al-Qaeda members after the US invasion of Afghanistan. The most famous of the al-Qaeda fugitives was Abu Zubaydah, a top operational commander who was captured in March 2003 at an LeT safehouse in Faisalabad, along with 16 LeT operatives who Pakistani police later released (Harrison 2002). In September 2003, at a Karachi madrassa linked to LeT, Pakistani authorities captured a group of students from Indonesia and Malaysia, including the brother of the mastermind behind the 2002 Bali bombing (Tankel 2011a, b).

The Headley case may be a harbinger of a new trend in which LeT operatives begin to link with other groups to carry out operations, sometimes operating independently. After his work scouting targets in India, Headley, on his own initiative, made contact with al-Qaeda and Ilyas Kashmiri, who further encouraged him to begin scouting targets in Copenhagen (Riedel 2012).

2.6 Conclusions

Even before the November 2008 Mumbai attacks, LeT was the subject of serious scrutiny by academics and journalists. However, important information about the organization's operations and decision-making remains opaque. Much of the information that is obtained is through LeT itself, which has a sophisticated media arm and tries to manage its public image. More comprehensive data about the membership and finances of LeT would be welcome, for international security analysts the critical issue is developing better mechanisms for predicting major LeT attacks.

(Clarke 2010) argues that LeT has built extensive networks of its own, and thus is no longer limited by the Pakistani government's ambitions and is increasingly pursuing an independent global agenda. Clarke concludes:

Although LeT was a key component of Islamabad's regional strategy in the past, the organization is growing beyond Pakistan's control and is undertaking its own independent operations. ...LeT has forged selective partnerships with fellow Pakistani and other militant groups, as well as criminal syndicates, whose activities undermine Pakistan's own security, escalate terrorism campaigns throughout South Asia, and increase the risk of inadvertent war between India and Pakistan. ...it appears that LeT leaders no longer feel that they are accountable to their former patron as a whole, but rather to themselves and a few select officers in ISI and the Pakistani military (current and/or retired). However, if support for LeT from the Pakistani intelligence and military establishment continues unabated, LeT will become a multinational organization that determines its own agenda as it will have a wide range of sponsors and sources of funding, and will have fighters and other vital resources spread throughout several regions (Clarke 2010).

Stephen Tankel is also concerned about expanded LeT terror regionally and internationally as well as the Pakistani state's long-term decreasing control over LeT. However, he argues that LeT's social services network and public presence has been an important source of leverage for the Pakistani government in reining in LeT:

...Lashkar controls a robust infrastructure and operates in plain sight.... This freedom of movement carries with it a number of benefits... but also serves as a leverage point that can be used to constrain Lashkar's activity.

However, Tankel is not sanguine about LeT's potential for international or expanded regional terror. He states, "The leadership's ongoing relationship with the ISI and the susceptibility to state pressure robs it of legitimacy in the eyes of some jihadis, who respect the sacrifices al-Qaeda leaders have made and the forthright manner in which they challenge the US as well as its many allies." Tankel argues that this creates tension within the organization to prove its commitment to jihad. Tankel states that this tension may have led LeT to launch the November 2008 assault on Mumbai, begin scouting targets in Denmark, and expand its operations in Afghanistan. Further, LeT (or LeT operatives working on their own) could render assistance to other terrorist groups, while maintaining ambiguity about its own role (Tankel 2011a, b).

Policy responses to constraining LeT are limited. In his conclusions, Tankel discusses the possibility of inducing LeT to abandon violence. He argues that settling the Kashmir issue to the satisfaction of Pakistan's leadership would be an important step to achieving this goal, but not sufficient to guarantee that LeT would abandon violence (Tankel 2011a, b). Other scholars studying the issue have discussed the possibility of weaning LeT militants away from violence by working with their extended families and providing alternative opportunities (Nawaz 2010). Increased counterterror cooperation between the US and India combined with pressing Pakistan to reduce its support for LeT may reduce LeT's capabilities. But many analysts believe that this is unlikely because LeT is simply too important to Pakistan's leadership (Tellis 2010) (Fair 2011).

What follows is an alternative approach to answering these questions using temporal probabilistic models of LeT's behavior. The models, as discussed in previous chapters, are based on data not only about LeT's behavior since the organization's founding, but also the circumstances (social, political, economic, military) surrounding those behaviors. The temporal probabilistic rules are not hand-crafted—they are learned automatically from the data in a manner that is provably guaranteed to satisfy various mathematical and statistical conditions. The temporal probabilistic rules can be used to provide predictions about likely group behaviors given a set of conditions, shed light on the drivers of LeT's actions, and possibly point to potential policies to mitigate the threat presented by LeT. While this is a new approach to understanding terrorist group behavior, major corporations already use such data analytic tools to better understand and serve their customers. Much of Google and Yahoo's revenues are based on automated ways to classify the behaviors of individuals who are using their service. Policy-makers responsible for critical national security functions should have comparable tools available.

This work, however, goes beyond the derivation of temporal probabilistic rules that provide a better understanding of LeT's behavior. Chapters 10 and 11 of this book develop methods to generate policies designed to rein in LeT. None of these policies is simple—each contains 17 or more "do's and don't's" that jointly constitute a policy. These policies also use the "big data analytics" technologies used by major corporations like Google, Yahoo, and Amazon. These companies try to identify ways that would induce a certain user behavior (e.g., clicking on an ad)—analagously; this project leverages big data analytics technology to identify ways to induce a certain behavior from LeT (e.g., reduction in different types of terrorist acts).

References

Abbas, A. (2002, January). In god we trust. *The Herald*.

Abbas, H. (2005). *Pakistan's drift into extremism: Allah, the army, and America's war on terror.* London: M.E Sharpe.

Abou Zahab, M., & Roy, O. (2004). *Islamist networks: The Afghan-Pakistan connection.* New York: Columbia University Press.

Abou Zahab, M. (2007). 'I shall be waiting for you at the door of paradise:' The Pakistani Martyrs of the Lashkar-e Taiba. In A. Rao, M. Bollig, & M. Bock (Eds.), *The practice of war: Production, reproduction and communication of armed violence* (pp. 117–126). New York: Berghahn Books.

Ahmad, T. (2012). Pakistani Jihadist Organization Jamaatud Dawa/Lashker-e-Taiba (LeT) Reappears on Internet, Promotes Antisemitism and Jihad, Leads Mass Protests Against American and India. *Inquiry & Analysis Report*, No. 791, Middle East Media Research Institute, February 2, 2012 from http://www.memri.org/report/en/0/0/0/0/0/0/6051.htm#_edn1

Al-Jazeera. (2010). Jammat chief rejects Indian charges. February 10, 2010. http://english.aljazeera.net/news/asia/2010/02/201021785121810598.html

Amir, R. (2003). Living with the faithful—a neighborhood choice. *Daily Times*, September 22, 2003. http://www.dailytimes.com.pk/default.asp?page=story_22-9-2003_pg7_9

ANImultimedia. (2011). Hafiz threatens to retaliate if India continues 'unprovoked firing. Uploaded by ANImultimedia. May 16, 2011. http://youtube/WnPipVa0600

Associated Press. (2004). Two 'Paintball' Terrorists Sentenced. April 9, 2004.

Associated Press. (2009). Former US university student convicted of terror charge. June 10, 2009.

Azzam Publications. (2001). Warning regarding British Muslims going for Jihad in Afghanistan. November 2001. http://www.azzam.com/afghan/html/afghanarticlebritishmuslims.htm

BBC News. (2007). Australia terror plotter' jailed. March 15, 2007.

BBC News. (2010). Pakistan 'gave funds' to group on UN terror blacklist. June 16, 2010. http://www.bbc.co.uk/news/10334914

Bearak, B. (2000a). A Kashmiri mystery. *New York Times Magazine*, December 31, 2000. http://www.nytimes.com/2000/12/31/magazine/a-kashmiri-mystery.html?pagewanted=all&src=pm

Bearak, B. (2000b). Gunmen Kill 3 at Garrison in New Delhi's Center. *New York Times*, December 23, 2000.

Bearak, B. (2000c). Lahore journal; A Jihad leader finds the U.S. Perplexingly Fickle. *The New York Times*, October 10, 2000.

Bergen, P. (2001). *Holy War Inc.: Inside the secret world of Osama Bin-Laden*. Free Press, New York.

Brenner, J. (2011). *America the vulnerable*. New York: The Penguin Press.

Bright, A. (2008). Pakistani organization accused of links to Mumbai attacks holds open house. *Christian Science Monitor*, December 5, 2008.

Brulliard, K. (2010). Afghan intelligence ties Pakistani group Lashkar-i-Taiba to Recent Kabul Attack. *Washington Post*, March 3, 2010.

Chalk, P., & Fair, C. C. (2002, October 17). Lashkar-e-Tayyiba leads the Kashmiri insurgency. *Jane's Intelligence Review, 14*(10), 1–5.

Clarke, R. (2010). *Lashkari-Taiba: The fallacy of subservient proxies and the future of Islamist terrorism in India*. Strategic Studies Institute, March 2010 http://www.strategicstudiesinstitute.army.mil/pdffiles/pub973.pdf

Cohen, S. P. (2004). *The idea of Pakistan*. Brookings, Washington, DC.

Coll, S. (2004). *Ghost Wars: The secret history of the CIA, Afghanistan, and Bin Laden, from the Soviet Invasion to September 10, 2001*. New York: Penguin Press.

Curtis, L. (2011). Moving past Mohali: What next for India and Pakistan? *Foreign Policy*, March 31, 2011 http://www.heritage.org/research/commentary/2011/03/moving-past-mohali-what-next-for-india-and-pakistan

Dalrymple, W. (1993). *City of Djinns: A year in Delhi*. New York: HarperCollins.

Emerson, S. (2002). *American Jihad: The terrorists living among US*. New York: The Free Press.

ExpressIndia.com. (2010). Siege ends at Lal Chowk; 2 militants gunned down. January 7, 2010. http://www.expressindia.com/latest-news/Siege-ends-at-Lal-Chowk-2-militants-gunned-down/564467/

Fair, C. C. (2007). Militant recruitment in Pakistan: A New Look at the Militancy-Madrasah connection. *Asia Policy 4*, 107–134.

Fair, C. C. (2009a). Antecedents and implications of the November 2008 Lashkar-e-Taiba (LeT) attack upon several targets in the Indian Mega-City of Mumbai. Santa Monica: Rand Corporation, March 11, 2009. http://www.rand.org/pubs/testimonies/CT320/

Fair, C. C. (2009b). Militant recruitment in Pakistan: Implications for al-Qaeda and other organizations. *Studies in Conflict and Terrorism, 27*, 489–504.

Fair, C. C. (2010). Students Islamic movement of India and the Indian Mujihideen: An assessment. *Asia Policy 9* (January 2010). http://www.nbr.org/publications/element.aspx?id=414

Fair, C. C. (2011). Lashkar-e-Taiba beyond Bin Laden: Enduring challenges for the region and the international community. *Testimony prepared for the U.S. Senate, Foreign Relations Committee*, May 24, 2011. http://home.comcast.net/~christine_fair/pubs/Fair_CT_May_24_2011.pdf

Fisk, R. (2010). Face to face with Pakistan's most wanted. *The Independent*, March 26, 2010. http://www.independent.co.uk/news/world/asia/face-to-face-with-pakistanrsquos-most-wanted-1928001.html

Ganguly, M. (2001). A banking system built for terrorism. *Time*, October 5, 2001. http://www.time.com/time/world/article/0,8599,178227,00.html

Ganguly, S. (1997). *The crisis in Kashmir: Portents of war*. Signs of Peace: Cambridge University Press and the Woodrow Wilson Center Press, Cambridge.

Gill, K. (1997). *Punjab: The knights of falsehood*. Har Anand Publications, India http://www.satp.org/satporgtp/publication/nightsoffalsehood/index.html

Government of India. (2008). Mumbai terrorist attacks: Dossier of evidence. *The Hindu*: Online Edition of India's National Newspaper. http://www.hindu.com/nic/dossier.htm

Government of Pakistan. (2011). Council of Islamic ideology. http://www.cii.gov.pk/

Government of Pakistan. (nd). *President of Pakistan, general Muhammad Zia ul-Haq-Interview to Foreign Media*, Vol. II. Government of Pakistan, Islamabad.

Habibullah, W. (2008). *My Kashmir: Conflicts and the prospects of enduring Peace*. U.S. Institute of Peace Press, Washington, DC.

Harrison, S. (2002). Corruption and extremism in Pakistan: Why Musharraf clings to power. *New York Times*, May 10, 2002.

Haqqani, H. (2005). *Pakistan: Between Mosque and Military*. Carnegie Endowment for International Peace, Washington, DC.

Indian Express. (1999). The Fidayen—faithful to the death. November 12, 1999. http://www.indianexpress.com/res/web/pIe/ie/daily/19991112/ige12049.html

International Crisis Group. (2003a). Kashmir: Learning from the past. International Crisis Group Asia Report no. 70. December 4, 2003. http://www.crisisgroup.org/ ~ /media/Files/asia/south-asia/kashmir/070_kashmir_learning_from_the_past.pdf

International Crisis Group. (2003b). Pakistan: The Mullahs and the Military. International Crisis Group Asia Report no. 49. March 20, 2003. http://www.crisisgroup.org/en/regions/asia/south-asia/pakistan/049-pakistan-the-mullahs-and-the-military.aspx

International Crisis Group. (2005). The State of Sectarianism in Pakistan. International Crisis Group Asia Report no. 95. April 18, 2005. http://www.crisisgroup.org/ ~ /media/Files/asia/south-asia/pakistan/095_the_state_of_sectarianism_in_pakistan.pdf

Investigative Project on Terrorism. (2008). Abdullah Azzam profile. July 9, 2008. http://www.investigativeproject.org/profile/103

Jamal, A. (2002). From Madrasa to School. *News*, December 15, 2002.

Jamwal, N. S. (2003). Terrorists' Modus Operandi in Jammu and Kashmir. *Strategic Analysis* 27(3), 382–403.

Jane's Islamic Affairs Analyst. (2009). Deobandi groups and Ahl-e-Hadith. May 21, 2009.

Jane's Terrorism and Security Monitor. (2008). Surge of the Insurgents. September 5, 2008.

John, W. (2011) Caliphate's Soldiers: The Lashkar-e-Tayyeba's Long War, Amaryllis and the Observer Research Foundation.

Joseph, J. (2002). Terrorists had entered temple, ready for long haul. *Rediff on the Net*, Rediff.com. September 26, 2002. http://www.rediff.com/news/2002/sep/25guj22.htm

Jost, P.M., & Sandhu, H.S. (2000, January). The hawala alternative remittance system and its role in money laundering. http://www.treasury.gov/resource-center/terrorist-illicit-finance/Documents/FinCEN-Hawala-rpt.pdf

Kalhana's Rajatarangini. (1900). *Chronicle of Kings of Kasmir*. (M. A. Stein, Trans.). London: Elibron Classics

Karmon, E. (2000). *Fatah and the popular front for the liberation of Palestine: International Terrorism Strategies (1968–1990)*. Herzliya, Israel: International Institute for Counter-Terrorism, November 25, 2000. http://212.150.54.123/articles/articledet.cfm?articleid=145

Kashmir Herald (2002). Hafiz Mohammed Saeed: Pakistan's heart of terror. July 2002.

King, G. (2004). *The most dangerous man in the world: Dawood Ibrahim*. New York: Chamberlain Bros.

Khan, A. (1970). *Raider in Kashmir*. Islamabad: National Book Foundation.

Kohlmann, E. (2000). *Regina v. Mohammed Ajmal Khan, Palvinder Singh, and Frzana Khan*. March 11, 2000. http://web.archive.org/web/20000311154008/http://www.dawacenter.com/introduction/introduction.html

Kohlmann, E. (2006). Expert witness: Synopsis of testimony. *Regina v. Mohammed Ajmal Khan, Palvinder Singh, and Frzana Khan*, Exhibit EK/1." Source: Snakes brook Crown Court, United Kingdom http://nefafoundation.org//file/ekletwitnessreport.pdf

Laskar, R. (2011). Zakiur Rehman Lakhvi secretly communicating with LeT cadres from prison. *Press Trust of India*, October 25, 2011. http://www.dnaindia.com/world/report_zakiur-rehman-lakhvi-secretly-communicating-with-let-cadres-from-prison_1603242

MEMRI Blog. (2010a). In Friday Sermon, Lashkar-e-Taiba Founder Calls for Promulgating Shari'a. *Roznama Express*, April 10, 2010. http://www.thememriblog.org/blog_personal/en/26224.htm

MEMRI Blog. (2010b). Lashkar-e-Taiba Founder: Internal conflicts in Muslim Nations have damaged Islamic Ummah. *Roznama Express*, July 6, 2010. http://www.thememriblog.org/blog_personal/en/28321.htm

MEMRI Blog. (2010c). Lashkar-e-Taiba's Charity Arm Jamaatud Dawa Attacks 'Jewish Company Gillette,' Religious Groups Force Gillette to Abandon Beard-Shaving Program In Karachi. April 1, 2010. http://www.thememriblog.org/blog_personal/en/26044.htm

Ministry of Home Affairs, Govt. of India, Department of Jammu & Kashmir Affairs. (2012). http://mha.nic.in/uniquepage.asp?id_pk=306

Mir, A. (2008a). JuD's Saeed distances himself from Lakhvi. *Daily News & Analysis*, December 21, 2008. http://www.dnaindia.com/world/report_juds-saeed-distances-himself-from-lakhvi_1215692

Mir, A. (2008b). Lashkar denies involvement in Mumbai carnage. *ME Transparent*, November 29, 2008. http://www.metransparent.com/spip.php?page=article&id_article=4914&lang=en

Mir, A. (2009a). JamaatulDaawa spokesman impersonates as Lashkar-e-Taiba spokesman. *ME. Transparent*, January 4, 2009 http://www.metransparent.com/spip.php?page=article&id_article=5134&lang=en

Mir, A. (2009b). The gates are open. *Outlook India*, January 19, 2009. http://www.outlookindia.com/article.aspx?239500

Nawaz, S. (2010). *Congressional Testimony on Islamist Militancy in Pakistan*. Washington: Atlantic Council March 11, 2010. http://acus.org/news/shuja-nawaz-congressional-testimony-islamist-militancy-pakistan

News. (2003). August 10.

Page, J. (2008). A monster out of control: Pakistan secret agents tell of militant links. *The Times of London*, December 22, 2008.

Pakistan Media Watch. (2010). Newspaper accepts paid advertisements from banned groups. September 7, 2010. http://pakistanmediawatch.com/2010/09/07/newspaper-accepts-paid-advertisements-from-banned-groups/

Pakistani Leaders Online. (2011). Prof. Hafiz Muhammed Saeed. http://www.pakistanileaders.com.pk/profile/Hafiz_Muhammad_Saeed

Parashar, S. (2012). Hafiz Saeed's brother-in-law Abdul Rehman Makki is a conduit between Lashkar-e-Taiba and Taliban. *Times of India*, April 5, 2012. http://timesofindia.indiatimes.com/world/pakistan/Hafiz-Saeeds-brother-in-law-Abdul-Rehman-Makki-is-a-conduit-between-Lashkar-e-Taiba-and-Taliban/articleshow/12539443.cms

Post, T., & Brand, J. (1992). Help from the Holy Warriors. *Newsweek*, October 5, 1992.

Press Trust of India. (2010a). Akshardham attack: Gujarat HC upholds death for 3. Gujarat, June 2, 2010. http://zeenews.india.com/news/gujarat/akshardham-attack-gujarat-hc-upholds-death-for-3_630632.html

Press Trust of India. (2010b). JuD holds anti-India jehadi rally in Pakistani Kashmir. New Delhi/ Srinagar, February 04, 2010. http://www.hindustantimes.com/JuD-holds-anti-India-jehadi-rally-in-Pakistani-Kashmir/H1-Article1-505117.aspx

Rajmohan, P. G. (2006, January). *Terrorist attack in Bangalore: A profile*. Institute of Peace and Conflict Studies Special Report 10. http://www.ipcs.org/pdf_file/issue/484403525IPCS-Special-Report-10.pdf

Raman, B. (1998). *Markaz Dawa al Irshad: Talibanisation of Nuclear Pakistan*. South Asian Analysis Group, July 26, 1998. http://www.saag.org/common/uploaded_files/paper6.html

Raman, B. (2000). Lashkar-e-Toiba: Its past, present and future. *South Asia Analysis Group*, Paper No. 175, December 25, 2000.

Rana, A. (2002). *Jihad Kashmir-wa-Afghanistan*. Lahore: Mashal Books.

Rana, A. (2004). Jamaatud Dawa splits. *Daily Times*, July 18, 2004 http://www.dailytimes.com.pk/default.asp?page=story_18-7-2004_pg7_20

Rana, M. A. (2006). *A to Z of Jihadi Organizations in Pakistan*. (S. Ansan, Trans.). Pakistan: Mashal Books http://www.desistore.com/jehadiorg.html

Rana, A. R. (2008, October–December) Jihadi Print media in Pakistan: An overview. *Conflict and Peace Studies, 1*(1).

Rediff.com. (1999) The encounter is still continuing. November 3, 1999 http://www.rediff.com/news/1999/nov/03kash.htm

Rotella, S. (2011a). Pakistan's terror connections, Chicago terrorism trial what we learned and what we didn't, about Pakistan's terror connections. *ProPublica* http://www.propublica.org/topic/mumbai-terror-attacks/

Riedel, B. (2012). *Deadly embrace: Pakistan, America, and the future of the global Jihad*. Washington, DC: Brookings Institution Press.

Rotella, S. (2011b). Pakistan and the Mumbai attacks: The Untold Story. January 26, 2011. *ProPublica* http://www.propublica.org/article/pakistan-and-the-mumbai-attacks-the-untold-story/

Roul, A. (2010a, April). After Pune, details emerge on the Karachi Project and its Threat to India. *CTC Sentinel, 3*(4).

Roul, A. (2010b). Jihad and Islamism in the Maldive Islands. Jamestown Foundation, February 12, 2010. http://www.jamestown.org/single/?no_cache=1&tx_ttnews%5Bswords%5D=8fd58 93941d69d0be3f378576261ae3e&tx_ttnews%5Bany_of_the_words%5D=Maldives&tx_tt news%5Btt_news%5D=36036&tx_ttnews%5BbackPid%5D=7&cHash=47c437e89488d099 18f9f2974eca684b

Roul, A. (2012, January). The Mastermind of Mayhem in Mumbai: A profile of Lashkar-e-Taiba's Zaki-ur-Rahman Lakhvi. *Militant Leadership Monitor* III, Issue 1.

Rubin, A. (2010). Militant group expands attacks in Afghanistan. *The New York Times*, June 15, 2010, http://www.nytimes.com/2010/06/16/world/asia/16lashkar.html?pagewanted=all

Sageman, M. (2004). *Understanding terror networks*. Philadelphia: University of Pennsylvania Press.

Sareen, S. (2005). *The Jihad factory: Pakistan's Islamic revolution in the making*. New Delhi: Har-Anand Publications.

Schofield, V. (1996). *Kashmir in the crossfire*. London: I.B Tauris.

Shah, S. (2010). Pakistan floods: Jamaat-ud-Dawa, Islamists linked to India's Mumbai attack, offer aid. *Christian Science Monitor*, August 4, 2010.

Siddiqi, K. (2000). Muridke complex: A nursery for Taiba men. *The Dawn*, May 7, 2000.

Sikand, Y. (2006). The SIMI story. *Countercurrents.org*, July 15, 2006. http://www.counter currents.org/comm-sikand150706.htm

Sikand, Y. (2007). Islamist Militancy in Kashmir: The case of Lashkar-e-Taiba. In A. Rao, M. Bollig, & M. Bock (Eds.), *The practice of war: Production, reproduction and communication of armed violence* (pp. 215–237). London: Berghahn Books.

South Asia Terrorism Portal. http://www.satp.org/satporgtp/countries/india/states/jandk/terrorist _outfits/lashkar_e_toiba_lt.htm

South Asian Terrorism Portal. (2011). Jaish-e-Mohammed (Army of the Prophet). Institute for Conflict Management, New Delhi. http://www.satp.org/satporgtp/countries/india/states/jandk/ terrorist_outfits/jaish_e_mohammad_mujahideen_e_tanzeem.htm

Stern, J. (2003). *Terror in the name of god: Why religious Militants kill*. New York: HarperCollins.

Swami, P. (1999, November 13–26). The growing toll. *Frontline, 16*(24). http://www.hinduonnet.com/fline/fl1624/16240390.htm

Swami, P. (2000a, May 13–26). The 'liberation' of Hyderabad. *Frontline, 17*(10). http://www.hindu.com/fline/fl1710/17100390.htm

Swami, P. (2000b, August 19–September 1). Turning the snow red. *Frontline, 17*(17). http://www.hindu.com/fline/fl1717/17170160.htm

Swami, P. (2004, May 9). Lashkar renews anti-American polemic. *The Hindu* http://www.hindu.com/2004/05/09/stories/2004050902141000.htm

Swami, P. (2005, November 17). Quake came as a boon for Lashkar leadership. *The Hindu* http://www.hindu.com/2005/11/17/stories/2005111705951200.htm

Swami, P. (2007). *India, Pakistan and the Secret Jihad: The Covert War in Kashmir, 1947–2004.* Abindon: Routledge.

Swami, P. (2008a, December 9). Pakistan and the Lashkar's Jihad in India. *The Hindu.*

Swami, P. (2008b, March 29–April 11). Tussle within. *Frontline 25*(7). http://www.hindu.com/fline/fl2507/stories/20080411250708700.htm

Tankel, S. (2011a). *Lashkar-e-Taiba: Past operations and future prospects.* National Security Studies program policy paper. New America Foundation: Washington, D.C. April 2011.

Tankel, S. (2011b). *Storming the world stage: The story of Lashkar-e-Taiba.* London: C. Hurst & Co.

Tehelka Magazine. (2009, October 17). The Professor of terror. *6*(41). http://www.tehelka.com/story_main43.asp?filename=Ne171009coverstory.asp

Tellis, A. (2010). Testimony by Ashley J. Tellis. *Bad Campany-Lashkar-e-Tayyiba and the Growing Ambition of Islamist Militancy in Pakistan.* Washington, DC: United States House of Representatives, Committee on Foreign Affairs, Subcommittee on Middle East and South Asia, March 11, 2010. http://carnegieendowment.org/files/0311_testimony_tellis.pdf

The 9/11 Commission Report. (2004).

The Daily Excelsior. (2009, February 1). 3 top LeT militants shot dead in Handwara. http://www.jammu-kashmir.com/archives/archives2009/kashmir20090201c.html

The Daily Times. (2009). Jaish-e-Muhammad builds huge base in Bahawalpur. September 14, 2009. http://www.dailytimes.com.pk/default.asp?page=2009\09\14\story_14-9-2009_pg7_16

The Dawn. (2010, November 24). Banned outfits raise cash from sacrifice day. http://www.dawn.com/2010/11/24/banned-outfits-raise-cash-from-sacrifice-day.html

The Economic Times. (2008, December 3). Captured Terrorist: Ajmal Amir Kasav tells his story. http://articles.economictimes.indiatimes.com/2008-12-03/news/27727691_1_afzal-ajmal-amir-kasav-faridkot-village

The Economic Times. (2010). Jamaatud-Dawa collects funds for jihad against India & US from PoK mosques. February 5, 2010. http://articles.economictimes.indiatimes.com/2010-02-05/news/28435559_1_pok-abdul-rehman-makki-jihad

The Guardian. (2005, July 18). Pakistan militants linked to London attacks. 2005.

The Hindu. (2000). Suicide squad storms Red Fort, kills 3 jawans. December 23, 2000. http://www.hindu.com/2000/12/23/stories/01230000.htm

The Times of India. (2003, August 8). Lashkar, Harkat issue fresh threats. .

The Times of India. (2009a). Pakistan's Jamaat 'ban' lie nailed. January 12http://articles.timesofindia.indiatimes.com/2009-01-12/pakistan/28028747_1_rally-lahore-channel

The Times of India. (2009b). Saeed under 'house arrest,' was Pak army's iftar guest. September 22, 2009. http://articles.timesofindia.indiatimes.com/2009-09-22/pakistan/28095513_1_saeed-under-house-arrest-lahore-ssp-sohail-sukhera-count-in-army-postings

The Times of India. (2012) LeT to Valley sarpanchs: Quit job or face music. April 17, 2012. http://articles.timesofindia.indiatimes.com/2012-04-17/india/31355272_1_sarpanchs-kashmir-issue-kashmiri-separatists

TV Gujarat. (2010). India's Most wanted HafeezSaeed rally in Karachi. Uploaded by TV Gujarat. June 16, 2010. http://youtube/gOPcowStOgc

U.S. Dept. of State. (2011). Office of the Coordinator for Counterterrorism, *Individuals and Entities Designated by the State Department Under E.O. 13224.* Washington, DC. August 16, 2011. http://www.state.gov/s/ct/rls/other/des/143210.htm

See also Office of Foreign Assets Control, U.S. Department of Treasury, "Changes to list of specially designated nationals and blocked persons since January 1, 2001. Washington, D.C.

U.S. Department of Treasury, October 13, 2011 http://www.treasury.gov/resource-center/sanctions/SDN-List/Documents/sdnew01.pdf

U.S. Dept. of Treasury (2008) Treasury targets LET leadership. May 27, 2008. http://www.treasury.gov/press-center/press-releases/Pages/hp996.aspx

U.S. Dept. of Treasury (2010) Resource Center: Hawala and alternative Remittance systems. December 3, 2010. http://www.treasury.gov/resource-center/terrorist-illicit-finance/Pages/Hawala-and-Alternatives.aspx

U.S. Dept. of Treasury. (2011a). Treasury sanctions Lashkar-e Tayyiba leaders and founders. September 28, 2011.http://www.treasury.gov/press-center/press-releases/Pages/tg1313.aspx

U.S. Dept. of Treasury. (2011b). Resource Center: Terrorism and Illicit Finance (last updated October 5, 2011). http://www.treasury.gov/resource-center/terrorist-illicit-finance/Pages/default.aspx

U.S. Dept. of Treasury. Protecting Charitable Organizations (Washington, D.C.: Department of Treasury). http://www.ustreas.gov/offices/enforcement/key-issues/protecting/charities_execorder_13224-a.shtml#ahpak

U.S. Dept. of Treasury. Protecting Charitable Organizations (Washington, D.C.: Department of Treasury). http://www.treasury.gov/resource-center/terrorist-illicit-finance/Terrorist-Finance-Tracking/Pages/fto_aliases.aspx

Waldman, A. (2003). Pakistan Criticized as India Bombing Toll Rises to 52. *The New York Times* http://www.nytimes.com/2003/08/27/world/pakistan-criticized-as-india-bombing-toll-rises-to-52.html

Waraich, O. (2009, May 13). Terrorism-Linked Charity Finds New Life Amid Pakistan Refugee Crisis. *Time* http://www.time.com/time/world/article/0,8599,1898127,00.html

Weinbaum, M., & Harder, J. (2008, March). Pakistan's Afghan policies and their consequences. *Contemporary South Asia, 16*(1), 25–38.

Yousef, M., & Adkin, M. (1992). *The Bear's trap: Afghanistan's Untold Story*. London: L. Cooper.

We are Ahlehadith. (2011). Hafiz Saeed Addressing Kashmir Rally Lahore 05 Feb 2011. http://aslaaf.webs.com/apps/videos/videos/show/12489100-video-speech-of-hafiz-muhammad-saeed

Chapter 3
Temporal Probabilistic Behavior Rules

Abstract This chapter describes the syntax and semantics of Temporal Probabilistic (or TP) behavioral rules used throughout the book to describe the behavior of Lashkar-e-Taiba. The chapter describes the intuition behind TP-rules, their formal syntax and meaning, and describes an algorithm used to derive the TP-rules automatically from data about LeT.

This chapter focuses on some of the technology underlying this book. Readers whose sole interest is in behavioral models of LeT and in policies to rein in violence carried out by LeT may skip this chapter without any loss of relevant material.

The first mathematical models of behaviors of terror groups were expressed via "Stochastic Opponent Modeling Agent" rules (or SOMA-rules for short) (Khuller et al. 2007) (Simari et al. 2012). A SOMA-rule is an expression of the form:

When condition C is true in the environment in which terror group G operates, there is a probability of p % that terror group G will take action A at intensity level I.

SOMA-rules have been used extensively to learn automatically expressed rules about the behaviors of terror groups such as Hezbollah (Mannes et al. 2008b) (Mannes and Subrahmanian 2009), Hamas (Mannes 2008a) and LeT (Mannes et al. 2011) and have been shown to have considerable expressive and predictive power. As an excellent example (Mannes et al. 2008b) predicted, in April 2008, Hezbollah's behavior in the first part of 2009 prior to scheduled Lebanese elections. In November 2008, a reporter for the Beirut *Daily Star* made skeptical comments on the predictions and included comments from Hezbollah about the predictions. So did a Lebanese academic political scientist. Hezbollah then behaved exactly as predicted in the first half of 2009 despite knowing about the predictions of (Mannes 2008b).

V. S. Subrahmanian et al., *Computational Analysis of Terrorist Groups: Lashkar-e-Taiba*,   69
DOI: 10.1007/978-1-4614-4769-6_3, © Springer Science+Business Media New York 2013

However, SOMA-rules can occasionally be a bit strange. An example of such a SOMA-rule describing the behavior of LeT that we derived automatically is given below.

LeT has armed clashes with local security forces with 87.5 % probability during a given month if:

- 0–4 commanders of LeT died during that month.
- the LeT diaspora provided support during that month.

Support = 14
Inverse Probability = 1
Negative Probability = 0

This SOMA rule is easy to read and understand. However, the reader is left confused about "cause and effect". As both the events "LeT has armed clashes with local security forces" and "0–4 commanders of LeT died" occurred during the same month, we are unable to tell whether the 0–4 LeT members died during the armed clashes with local security forces referenced in this rule, or whether they died independently of the armed clashes. The rule would be significant and interesting in the latter case, but totally uninteresting in the former case. The flaw in SOMA-rules is that it is often *hard to tell* which of the two cases is applicable. When the events described in the "if" part of the SOMA-rule clearly do not overlap with the events occurring in the "then" part of the same rule, we can clearly understand a significant correlative relationship between the "if" and then "then" parts of the rule. But when there is overlap—as in the case described above—it is difficult to tell.

A further flaw with SOMA-rules is that they do not include a temporal element. In the real-world, terror groups do not always react instantaneously to events in their environment. They deliberate internally, they develop operational plans, they plan logistics to support their plans, and they finally execute (or sometimes abort) their plans. All of this takes time, leading to temporal delays that SOMA-rules do not model.

This chapter, describes *temporal-probabilistic rules* (or TP-rules for short) introduced by (Dekhtyar et al. 1999) that avoid both these problems. Clearly articulating *delays* between the time when a terror group's environment satisfied a given condition, and a *subsequent* time (this book examines 1–3 month delays, though of course, longer delays are also possible and can/should be studied) when the terror group took a given action, cleanly excludes the possibility that the "if" part of a rule and the "then" part of the rule could have co-occurred at the same time.

This chapter starts by first briefly describing the structure of the data we have about LeT, followed by a description of the syntax of TP-rules, together with a description of the algorithm used to derive TP-rules.

3.1 Database Schema

All of the data about LeT is represented as a relational database table whose rows correspond to *months* and whose columns correspond to two kinds of *attributes*. In fact, the exact representation used in our CMOT project (Shakarian 2012) can be and has been used to represent information about a host of terror groups, not just LeT. Other terror groups for which data has been collected using the same representation include the Pakistani group Jaish-e-Mohammed (responsible for the attack on India's parliament in December 2001), the Indian Mujahideen, the Student Islamic Movement of India, and the Forces Democratiques de Liberation du Rwanda (FDLR) based in the Democratic Republic of Congo. The FDLR's leaders are widely held responsible for the 1994 genocide of Tutsi civilians in Rwanda.

Environmental *attributes* refer to aspects of the environment within which LeT functioned. These include attributes relating to:

- The internal structure of LeT;
- Information about the internal activities of LeT (e.g., inter-organizational conflict);
- Information about LeT's resources including financial, military, and political support, as well as similar kinds of support by the diaspora and foreign states;
- Information about LeT's ongoing campaigns (e.g., news/media campaigns);
- Information about LeT's facilities (e.g., training camps);
- Information about LeT's grievances (e.g., territorial claims, attempts to change religious practices);
- Information about actions (both supportive and adversarial) taken by external actors such as the US, India, and Pakistan towards LeT including bans of LeT, kills/arrests of LeT personnel, and issuance of arrest warrants or asset freezes.

The second class of attributes in modeling the behavior of LeT are *action attributes* describing the intensity of various types of actions taken by LeT. Actions are typically described by location (national or transnational), type of action (e.g., fedayeen attack vs. hijacking vs. armed clash), and type of target (e.g., security installation vs. transportation facility). This work studied numerous action attributes including:

- Armed attacks
- Suicide attacks
- Hijackings
- Abductions/kidnappings
- Attacks on security targets (e.g., police stations or military bases)
- Attacks on transportation targets (e.g., airports, train stations)
- Attacks against civilians on the basis of ethnicity or religion (e.g., against Hindus or Christians or Jews)
- Attacks on government facilities (including government offices) not related to security forces
- Attacks on symbolic or tourist sites (e.g., the Red Fort in Delhi)

- Armed clashes with different types of security forces.

As in standard relational databases, each attribute (environmental or action) has a precisely stated *domain*. In most work related to the modeling of ethnic groups or terror groups, the domains are binary (or 0-1) meaning the group either performed a given action or not.

In contrast, our LeT study tried to capture numeric information as best as possible. How many LeT leaders were arrested during a given month (to the best estimable extent possible)? How many armed clashes occurred during a given month between LeT and security forces? As a consequence, our data can capture variations in intensity—something that cannot be captured seamlessly by past efforts to model terror groups.

Appendix A provides a much more detailed description of both our data and our data collection methodology.

3.2 TP-Rule Syntax

TP-rules use a form of logic programming and computational logic to express rules that have both a temporal and a probabilistic aspect. TP-rules are a variant of Generalized Annotated Programs (Kifer and Subrahmanian 1992) that were first introduced in (Dekhtyar et al. 1999) and later studied and extended to *annotated probabilistic temporal* (APT) logic programs in (Shakarian et al. 2011a, b).

Every attribute p in the data set described in Sect. 3.1 corresponds to a *unary predicate symbol*. The argument to such a predicate symbol can either be a value v from the domain of the attribute p or a variable X ranging over this domain. v and X are called *terms* over the domain of p.

If t is a term over the domain of predicate symbol p, then $p(t)$ is called an *atom*. When t is in the domain of p, $p(t)$ is called a *ground atom*. When p is an action (resp. environmental attribute), $p(t)$ is an *action atom* (resp. *environmental atom*).

For example, suppose *suicide-attack* is an action attribute whose domain is the non-negative integers (0 and up). Then *suicide-attack*(3) is a ground atom referring to 3 suicide attacks. Likewise, *suicide-attack*(10) is a ground atom referring to 10 suicide attacks, while *suicide-attack*(X) is a ground atom that can be instantiated to any number of suicide attacks.

Suppose $t' \geq 0$ is either a time point (non-negative integer) or a variable ranging over non-negative integers, and $p' \varepsilon [0,1]$ is a probability. Then $[t',p']$ is called a *temporal probabilistic annotation* (or *tp-annotation* for short).

For example, [10,0.7] refers to a probability of 70 % or more of some (unspecified event) occurring 10 time units after a given time.

If $p(t)$ is an atom and $[t',p]$ is a tp-annotation, then $p(t){:}[t',p']$ is a *temporal probabilistic annotated atom* (*tp-annotated atom*, for short).

For instance, *suicide-attack* (3): [2,0.8] is a tp-annotated atom that says that 3 suicide attacks will occur with 80 % probability 2 time units after some specified time.

If $p(t)$ is an (action, environmental) atom and $[t',p']$ is a tp-annotation, then $p(t):[t',p']$ is a (resp. action, environmental) *tp-annotated atom*. $p(t):[t',p']$ is *ground* if and only if $p(t)$ is ground—otherwise it is *non-ground*.

If X is a variable over the non-negative integers and Y is a term over the non-negative integers, then $X = Y$, $X \leq Y$, $X < Y$, $X \geq Y$ are all comparison atoms.

If $A_1,..,A_n$ are environmental atoms or comparison atoms and $p(t):[t',p']$ is a tp-annotated action atom, then

$$p(t) : [t',p'] \leftarrow A_1 \& \ldots \& A_n$$

is a *temporal-probabilistic rule* (TP-rule for short). Intuitively, this rule says that if $A_1,..,A_n$ are all true of LeT's environment at a given time τ, then $p(t)$ is true at time $(\tau + t')$ with probability p'. $p(t)$ is called the *head* of this TP-rule, while $A_1 \& \ldots \& A_n$ is called the *body*.

For instance, the TP-rule

TP-Rule AAH-1

- LeT attacks civilians one month after months in which:
- LeT was a religious organization and
- Between 0 and 2 LeT commanders died.

Support = 10
Probability = 100 %, *Inverse Probability* = 100 %, *Negative Probability* = 0 %

---

which we discuss in further detail in Chap. 4 can be written in TP-rule syntax as:

$$civilian\text{-}attack(1) : [1,1] \leftarrow religious(1) \,\&\, leaders\text{-}died(X) \,\&\, X \leq 2.$$

This rule says that LeT carries out one civilian attack with 100 % probability one month after months in which X leaders died and X was 2 or less in number.

Note that there is also data in our LeT data set that looks not at the number of events, but uses a number to denote the intensity of events on a qualitative scale.

Another example of a TP-rule that we derived about attacks by LeT against security forces is given in Chap. 6 and says:

TP-Rule PSF-2

- LeT carries out 1 attack against professional security forces two months after months in which:
- There was no government ban on LeT and
- LeT had a strong relationship with the Pakistani military.

Support = 13
Probability = 92.9 %, *Inverse Probability* = 92.9 %, *Negative Probability* = 33.3 %

This TP-rule can be formally expressed as:

$$profsecforce\text{-}attack(1) : [2, 0.929] \leftarrow govt\text{-}ban(0) \,\&\, military\text{-}relationship(1).$$

This rule uses qualitative information in the TP-rule body and quantitative information in the rule head. It says that when the level of a government ban on LeT is 0 (indicating no ban) and when the level of a relationship between LeT and the Pakistani military is 1 (indicating existence of the relationship), then LeT carries out one attack against a professional security force 2 months later with 92.9 % probability.

In this book, we do not go into the details of the logical and probabilistic semantics of TP-rules as the goal of the book is to study how to model the behavior of LeT using TP-rules, nor do we go into the details of the algorithms used to reason with the TP-rules. Such details are presented in detail in (Dekhtyar et al. 1999) and (Shakarian et al. 2011a, b).

3.3 SOMA Rules

Subrahmanian and Ernst (Subrahmanian and Ernst 2009) have developed methods to automatically learn "SOMA-rules" *without time* from data of the kind described in Sect. 3.1. SOMA-rules are probabilistic rules that use the syntax of probabilistic logic programs (Ng and Subrahmanian 1993).

As SOMA-rules are not used extensively in this book, we review them very briefly here. An annotated atom is an expression of the form $A: [v_1, v_2]$ where v_1, v_2 are reals in the [0,1] interval. If $A_1, .., A_n$ are environmental atoms or comparison atoms and $p(t):[v_1, v_2]$ is an annotated atom, then

$$p(t) : [v_1, v_2] \leftarrow A_1 \,\&\, \ldots \,\&\, A_n$$

is a SOMA-rule. Intuitively, this rule says that if $A_1 \,\&\, \ldots \,\&\, A_n$ are all true, then $p(t)$ is true with probability in the interval $[v_1, v_2]$. $p(t)$ is called the *head* of this SOMA-rule, while $A_1 \,\&\, \ldots \,\&\, A_n$ is called the *body*.

SOMA-rules have been used extensively to reason about the behavior of terror groups such as Hezbollah, Hamas, and Lashkar-e-Taiba. In addition, SOMA-rules have been extracted automatically from data about a total of over 50 groups. In a completely different setting, SOMA-rules have been used to learn relationships between socio-political-economic data and educational outcomes in 221 countries.

3.4 Extracting SOMA-Rules Automatically

In this section, we briefly describe the method of (Subrahmanian and Ernst 2009) to extract SOMA-rules automatically. This method has been extended in a straightforward manner to automatically extract TP-rules—the extraction of TP-rules is described in the next section.

The Subrahmanian-Ernst (SE for short) algorithm generates rules whose bodies consist of *bi-conjuncts* which are expressions of the form $p(X) \ \& \ L \le X \le U$. A *bi*-SOMA rule is an expression of the form

$$p(t) : [v_1, v_2] \leftarrow B_1 \ \&. \ . \ .\& \ B_n$$

where each B_i is a *bi*-conjunct. We call $B_1 \ \&...\& \ B_n$ a *bi*-body.

It is easy to see that each *bi*-SOMA rule is also an ordinary SOMA-rule. The SE algorithm generates *bi*-SOMA rules automatically.

Given two *bi*-bodies C_1 and C_2 of environmental atoms, we say that C_1 and C_2 are *equivalent* (with respect to a given data set such as our LeT data set) if and only if two conditions hold:

1. the set of months in which C_1 is true for LeT is identical to the set of months in which C_2 is true, i.e. the two conditions always co-occurred and
2. C_1 and C_2 involve exactly the same environmental atoms.

The equivalence relation above induces equivalence classes on the set of all *bi*-bodies. From each equivalence class, a *canonical member* is selected to represent that equivalence class. As all *bi*-bodies within an equivalence class are equivalent in terms of the environmental attributes they reference and in terms of when they are true, any member of the class is representative of the class as a whole.

Canonical members are required to be *tight*. We don't plan to define tight formally here, but instead give an informal example. Suppose an equivalence class consists of the *bi*-bodies $C_1, C_2,.., C_n$. Suppose C_i contains the *bi*-conjunct $p(X) \ \& \ L_i \le X \le U_i$. Clearly all the C_i's must contain such *bi*-conjuncts (otherwise they would not be in the same equivalence class). Then the canonical representative of the class must contain the *bi*-conjunct

$$p(X) \& \ min\{L_i | i = 1, ..n\} \le X \le max\{U_i | i = 1, .., n\}.$$

The same principle applies to all other attributes occurring in the equivalence class $\{C_1, C_2,.., C_n\}$.

The *dimension* of a *bi*-body is the number of attributes in it. The SE algorithm defines a "simpler-than" ordering on *bi*-bodies as follows.

C_1 is "simpler than" C_2, denoted $C_1 \gg C_2$ iff C_1 has the same number (or fewer) attributes in it than C_2.

Intuitively, the "simpler than" relationship merely looks at the number of attributes in *bi*-bodies. However, this is not enough.

Given a specific action variable value $p(t)$ that we want to predict, the SE algorithm then defines a more general ordering on conditions. $C_1 \gg C_2$ iff:

$$C_1 \gg C_2 \text{ and}$$
$$\text{Conf}(C_1) \geq \text{Conf}(C_2) \text{ and}$$
$$\text{Sup}(C_1) \geq \text{Sup}(C_2)$$

where Conf(C) is the confidence of condition C w.r.t. predicting $p(t)$, i.e.,

$$Conf(C) = \frac{number\ of\ months\ when\ C\ was\ true\ and\ p(t)was\ true}{number\ of\ months\ where\ C\ was\ true}$$

and Sup(C) is the support of C which is the numerator of the above formula.

Intuitively, for a *bi*-body C_1 to be better than a bi-body C_2 (i.e., for $C_1 \gg C_2$ to hold), we require that C_1 not only be simpler than C_2 but also that C_1 have higher support and confidence than C_2.

Throughout this chapter, we assume that we only consider *bi*-bodies of dimension p or less for some specified p. Intuitively, p denotes how large a *bi*-body can be and can be set by the system. In our rule extractor, we set p to 3 which means that the body of any SOMA or TP-rule we consider will have at most 3 *bi*-conjuncts in it.

Given a *bi*-body C, the *up-set* of C, denoted $up(C)$, is the set of all *bi*-bodies of dimension p or less such that C' is simpler than C. Formally, $up(C) = \{C' \mid C'$ is a tight *bi*-body and C' is of dimension p or less and $C' \gg C\}$.

We can now define a way of iteratively computing such *bi*-bodies.

$Tp\!\uparrow\!1 = \{C \mid C$ is tight and dimension(C) $\leq p$ and up(C) = $\{\}\}$.
$Tp\!\uparrow\!(i + 1) = \{C \mid C$ is tight and dimension(C) $\leq p$ and up(C) is a subset (or equal to) $Tp\!\uparrow\!i\}$.

When computing $Tp\!\uparrow\!k$, the SE algorithm uses something called a *condition graph* (COG) whose vertices are (labeled with) tight *bi*-bodies. There is an edge from vertex u to vertex v if :

1. $u.BiBody \gg v.BiBody$ and
2. There is no vertex w such that $u.BiBody \gg w.BiBody \gg v.BiBody$.

Each vertex in a COG has a level. A vertex with in-degree 0 has level 0; otherwise, the level of a vertex v is $1 + \max\{$level(u) | there is an edge from u to v in the COG$\}$.

Rather than building a COG completely, the SE algorithm uses COGs to generate $Tp\!\uparrow\!k$ efficiently when a specific outcome (e.g., *civilian-attack*(1)) is specified as input, i.e., we want to find rules that effectively predict when LeT launches attacks against civilians with intensity 1.

The following algorithm contains several procedures that are briefly summarized below.

- The **Build-Data-Struc**() procedure builds a data structure that contains all rows in the LeT database that satisfy the desired Outcome condition (in the case of the

SatisfyOutcome line) or do not satisfy the desired Outcome condition (in the case of the notSatisfyOutcome line).

- The **GenerateTightBi-bodies**() function generates all tight *bi*-bodies associated with attributes in φ using the table DB containing the LeT data. For each such tight bi-body v generated, it creates a record with the fields *bibody* (specifying the *bi*-body), *conf* (specifying the confidence of the rule *Outcome* $\leftarrow$ *v*) and *sup* (specifying the support).
- The **InsertCOG** function inserts the vertex v into the COG if the level of the vertex would be K or less. It then updates the neighbors of the vertex so that their levels are appropriately reset.

The **Compute Tp↑k** algorithm finds all *bi*-bodies in Tp↑k and these *bi*-bodies form the bodies of rules whose head is *Outcome*. By invoking this function for all desired action attributes (and associated values) associated with LeT, we were able to derive all possible rules that satisfied desired support and confidence levels.

```
Compute Tp↑k(DB, ENV, Outcome, p, k)
COG = NIL; (* initially there are no vertices in the COG *)
Foreach combination φ of p or less environmental attributes do
SatisfyOutcome = Build-Data-Struc(DB, ENV, φ, Outcome);
NotSatOutcome = Build-Data-Struc(DB, ENV, φ, ~Outcome);
(* SatisfyOutcome contains the projection of DB on attributes in φ for
months that satisfy the desired outcome; NotSatOutcome does the same for
months that do not *)
TightBi-Bodies = GenerateTightBi-Bodies (φ, SatisfyOutcome);

(* generates tight bi-bodies associated with attributes in φ and
computes their support *)
Foreach v in TightBi-Bodies do
(* v is record with fields bibody, conf,sup *)
numNotOutcome=
CountQuery(v.bibody, NotSatisfyOutcome)

(* finds number of months in which v.bibody was true
but Outcome did not occur *)
v.confidence = v.support/(v.support + numNotOutcome);
COG = InsertCOG(v, COG, K);
endfor;
endfor;
return ExtractBiBody(COG)

end
```

3.5 Automatically Extracting TP-Rules

In order to automatically extract TP-rules associated with a time offset of j months, all we had to do was to invoke the **Compute Tp↑k** algorithm with a new database DB' (instead of the actual LeT database DB).

DB' can be calculated from DB as follows:

- Set $DB1$ to DB.
- Eliminate all rows in $DB1$ associated with the first j months.
- For each month m in $DB1$, replace all environment attribute values $m.E$ by the environmental attribute value $(m\text{-}j).E$ from the original table DB.
- DB' is the result.

What this manipulation does is "pretend" that the environment in month m is actually the environment in month $(m\text{-}j)$ and thus, it "fools" the **Compute Tp↑k** algorithm into computing the correct time-offset rules.

3.6 Conclusion

The system computed TP-rules for time offsets 1,2,3,4, and 5 months for the entire LeT dataset. However, most of the analysis reported in subsequent chapters focuses only on time offsets of 1–3 months.

Computing all TP-rules that satisfy various support and confidence criteria is challenging and expensive (from a compute time perspective). Nonetheless, we were able to generate over 15,000 TP-rules. Many of these rules were either uninteresting or repetitive—we manually went through all of these TP-rules in order to find the interesting ones.

Chapters 4–9 describe the most interesting TP-rules that we discovered pertaining to six types of "bad acts" by LeT:

- Attacks on civilians;
- Attacks against public sites, tourist sites, and transportation infrastructure;
- Attacks against professional security forces;
- Attacks against security installations;
- Attacks of other types including attacks on holidays, attacks on the government, and attempted (but unsuccessful) attacks, and
- Armed clashes.

References

Dekhtyar, A., Dekhtyar, M., Subrahmanian, V.S. (1999). Temporal Probabilistic Logic Programs: Proceedings 1999 International Conference on Logic Programming, November 1999. New Mexico. (pp. 109–123).

Khuller, S., Martinez, V., Nau, D., Simari, G., Sliva, A., & Subrahmanian, V. S. (2007). Computing most probable worlds of action probabilistic logic programs: scalable estimation for 1030,000 worlds. *Annals of Mathematics and Artificial Intelligence, 51*(2–4), 295–331.

Kifer, M., & Subrahmanian, V. S. (1992). Theory of Generalized Annotated Logic Programming and its Applications. *Journal of Logic Programming, 12*(4), 335–368.

Mannes, A., Subrahmanian, V.S. (2009). Calculated terror, *Foreign Policy magazine (online edition)*, December 15, 2009. Retrieved http://www.foreignpolicy.com/articles/2009/12/15/calculated_terror?page=full

Mannes, A., Sliva, A., Subrahmanian, V.S. (2011). *A computational enabled analysis of Lashkar-e-Taiba attacks in Jammu and Kashmir: Proceedings 2011 IEEE European intelligence & security informatics conference September 2011*. Athens.

Mannes, A., Sliva, A., Subrahmanian, V.S., Wilkenfeld, J. (2008a). *Stochastic opponent modeling agents: a case study with hamas: proceedings 2008 international. conference on computational cultural dynamics* (pp. 49–54), September 2008. AAAI Press.

Mannes, A., Michaell, M., Pate, A., Sliva, A., Subrahmanian, V.S., Wilkenfeld, J. (2008b). In Liu, H., Salerno, J., Rogers, M. (Eds.) *Stochastic opponent modelling agents: a case study with hezbollah: Proceedings 2008 first intl. workshop on social computing, behavioral modeling and prediction*, April 1–2, 2008, Springer Verlag, Phoenix.

Ng, R., Subrahmanian, V.S. (1993). Probabilistic Logic Programming, *Information and Computation, 101*, 2, 150—201.

Shakarian, J. (2012). The CMOT Codebook, available from the Lab for Computational Cultural Dynamics, University of Maryland Institute for Advanced Computer Studies, University of Maryland, College Park, MD 20742. Extended and revised by Schuetzle, B. and Nagel, M. in 2012.

Shakarian, P., Simari, G. I., & Subrahmanian, V. S. (2011a). Annotated probabilistic temporal logic: approximate fixpoint implementation. *ACM Transactions on Computational Logic, 12*, 1–42.

Shakarian, P., Parker, A., Simari, G., Subrahmanian, V.S. (2011b). Annotated probabilistic temporal logic. *ACM Transactions on Computational Logic, 12*(2).

Simari, G., Martinez, V., Sliva, A., Subrahmanian, V.S. (2012) Focused most probable world computations in probabilistic logic programs. *Annals of Mathematics and Artificial Intelligence.*, 64 (Nrs2-3):113–143.

Subrahmanian, V.S., Ernst, J. (2009). Method and system for optimal data diagnosis, US patent Nr. 7474987, January 6.

References

Chapter 4
Targeting Civilians

Abstract LeT has carried out numerous operations in which civilians were both the real and intended targets. This chapter focuses on the conditions under which LeT carried out attacks against civilians in general, or civilians on the basis of their religious identity (primarily Hindus, but also taken broadly to include Christians and Jews). The chapter discusses several TP-rules derived automatically from the LeT data set used in this book.

This chapter focuses on the circumstances under which LeT has carried out attacks against civilians in general and when LeT specifically targeted individuals on the basis of their religious beliefs.

Although LeT spokespeople deny that LeT targets civilians, there is little question that LeT frequently adopts this tactic. Most of the victims in these attacks have been Hindus. In some cases this is because Hindus represent the majority of India's population, but LeT rhetoric is frequently and rabidly anti-Hindu (Rana 2006)[1] and there are many cases in which LeT has followed through on its rhetoric by targeting Hindus specifically.

For example, on many occasions in the late 1990s and early 2000s LeT terrorists massacred Hindus in Kashmir in an attempt to fuel ethnic tensions (Tankel 2011a). In August 2002 when LeT gunmen attacked a group of pilgrims traveling to the Amarnath shrine in Kashmir, there was little ambiguity that the pilgrims were targeted specifically because they were Hindu (Ahmad 2002). In other cases, LeT gunmen singled out Hindu families or villages within Kashmir for massacres (United Press International 2002). There have been other cases in which LeT has targeted other religious groups, such as Jews and Sikhs. The LeT attack on the Nariman House, which housed a Jewish religious institution, in the November

---

[1] According to (Rana 2006, p. 322) LeT's magazine *Mujalla-ul-Dawa* reported an August 2001 rally on the Indian-Pakistan border in which LeT officials led the crowd in chants denigrating Hindus.

V. S. Subrahmanian et al., *Computational Analysis of Terrorist Groups: Lashkar-e-Taiba*, 81
DOI: 10.1007/978-1-4614-4769-6_4, © Springer Science+Business Media New York 2013

2008 siege of Mumbai was an example of targeting based strictly on religious identity.

The other types of TP-rules discussed in this chapter refer to attacks on civilians in general, in which the victims do not have a defining characteristic other then being civilians. In some cases, there was no mention made in the sources available of the victims' religious affiliation, and in other cases the nature of the attack was such that it was not clearly targeting a specific religious community. Setting off bombs in public places (but not temples) in India would be an attack on civilians in general but not targeting a specific religion. Finally, it is worth noting that this variable measured number of incidents. A grenade attack that took no lives and an attack by gunmen that took several lives would each count as one incident (National Counterterrorism Center's World Wide Incidents Tracking System).[2]

The investigation extracted a total of

- 99 TP-rules with a time offset of 1 month dealing with attacks targeting civilians,
- 8 TP-rules with a time offset of 2 months dealing with attacks targeting civilians,
- 61 TP-rules with a time offset of 3 months dealing with attacks targeting civilians.

This chapter does not analyze all of the TP-rules that our system derived. Following are summaries of some TP-rules that highlight key indicators. The variables affecting LeT attacks on civilians included the following:

- *Death of LeT Commanders*. Attacks followed within 1–3 months of the time when LeT field commanders died (usually in combat with Indian forces). Again, these correlations were conditioned on some other variables.
- *Intra-Organizational Conflict, Group Cohesion*. The lack of intra-organizational conflict and LeT's ability to avoid splintering into factions seem to precede attacks on civilians by 1–3 months.
- *Government Offensive Actions Against LeT*. Offensive government actions by Pakistan, India, and the Jammu and Kashmir provincial government such as bans on LeT, asset freezes against LeT or its leaders, raids directed towards LeT, or kills of LeT leadership cadres, seem to be followed—within 1–3 months—by attacks against civilians when certain other conditions are true.
- *Desertion by LeT members*. Within one month of desertion of LeT members, there appear to be attacks against Hindus.

The TP-rules discovered automatically by our data mining algorithm are correlative, not necessarily causative. It is not clear how to establish causal relationships between environmental factors and actions by LeT as these causal

---

[2] This is obviously an imperfect measure of terrorist activity, which has both strengths and weaknesses. A discussion of some of the challenges in quantitative measures of terrorist activity can be found on the Methodology section of the National Counterterrorism Center's World Wide Incidents Tracking System—wits.nctc.gov.

relationships require the ability to set up an experimental design that would allow the researcher to vary the environmental variables and observe the reactions in much the same way a physicist or a biologist modifies variables, makes observations, and draws conclusions. Unfortunately, this is infeasible to do—not only in the context of LeT—but in the context of most work on understanding terror groups in general.

In addition, it must be emphasized that the above list of variables is just a list—most of the TP-rules consist of complex combinations of the variables listed above. The rest of the chapter is a sampling of these TP-rules.

4.1 Attacks Against Civilians and Deaths of LeT Commanders

Irrespective of whether the results looked at one, two, or three-month offsets, deaths of LeT commanders was a leading indicator of attacks by LeT against civilians. The vast majority of LeT commanders covered in this variable are field commanders killed fighting in Jammu and Kashmir. March 2004, a not atypical month, saw the deaths of an LeT divisional commander, a district commander, and an area commander in clashes with Indian security forces in Jammu & Kashmir (South Asia Terrorism Portal 2011a).[3]

TP-Rule AAH-1. LeT attacks civilians one month after months in which:

- LeT was a religious organization and
- Between 0 and 2 LeT commanders died.

Support = 10
Probability = 100 %,   *Inverse*   *Probability* = 100 %,   *Negative*
Probability = 0 %

TP-rule (AAH-1) says that if 0–2 LeT commanders die in a given month *and* LeT was claimed to be a religious organization during that same month,[4] then there is a 100 % probability that they will carry out attacks on civilians in the following month.

The inverse probability specification says that in 100 % of all months in which LeT claimed to be a religious organization and in which 0–2 of their commanders died, it was the case that LeT carried out attacks against civilians during the following months.

---

[3] See the chronologies maintained by the South Asia Terrorism Portal—satp.org.
[4] LeT's claim to be a religious organization is true virtually for the entirety of LeT's life-span.

The negative probability of 0 for this TP-rule indicates that there was not a *single month* in which LeT carried out these attacks which has not preceded immediately by a month in which LeT both claimed to be a religious organization and 0–2 of LeT's commanders died.

Finally, the support of 10 indicates that there were 10 months in which three things happened: LeT made religious claims and 0–2 of their commanders died and LeT attacked civilians 1 month later.

The data shows therefore that attacks on civilians attributed to LeT are *perfectly aligned* with cases when LeT claimed to be a religious organization and lost 0–2 of its commanders in the preceding month.

TP-Rule (AAH-1) is *further supported* by additional rules derived with the exact same probability, inverse probability, negative probability, and support when 0–1 LeT commanders died in a month rather than 0–2 LeT commanders dying.

TP-Rule AAH-2. LeT carried out 1–9 attacks against Hindus three month after months in which:

- LeT made no territorial claims and
- Between 0 and 2 LeT commanders died.

Support = 10
Probability = 90.9 %,   *Inverse*   *Probability* = 90.9 %,   *Negative*
Probability = 33.3 %

TP-Rule (AAH-2) above shows a similar situation, except that the attacks explicitly targeted Hindus with a 3-month offset.

Unlike TP-Rule (AAH-1), this rule is less clear-cut, but still offers strong evidence for being valid. It has a high probability (90.9 %), a high inverse probability (also 90.9 %) and a 33.3 % negative probability.

The negative probability states that there is a 33.3 % probability that LeT will attack Hindus 3 months after a month in which either LeT made territorial claims or more than two of LeT's commanders died.

It is unclear *why* Rules (AAH-1) and (AAH-2) are true, but the conditions suggest some possibilities. One possibility is simply that massacres occur during periods of high LeT activity, which exacts a great cost on LeT's battlefield commanders, who are frequently killed in action. Another possibility is that when LeT's commanders are killed in a given month, a certain degree of dismay may set in amongst some of the cadres. Carrying out a successful attack is likely to reinvigorate the organization and energize lower-rung operatives and Hindu civilians are easy and accessible targets.

Section 4.2 will present additional rules that show that months when LeT's commanders are either killed or arrested or face arrest warrants or have asset freezes against them (or when LeT is banned by Pakistan) are followed by attacks on Hindus a few months later by LeT.

4.2 Attacks Against Hindus and Government Action Against LeT

We tracked a wide variety of government actions against LeT including those listed below. The governments involved in such actions include India, Pakistan, as well as specific actions by the provincial government of Jammu and Kashmir.

- *Arrests*. This variable refers to arrests of LeT personnel by government forces. Arrests occur under a wide variety of contexts. Sometimes, particularly in Kashmir, they are the product of an armed clash, when either an attempted arrest becomes an armed clash or LeT operatives are captured in an armed clash. In other cases, arrests occur either after major LeT operations or as a result of intelligence gleaned, such as in October 2003 when Indian police arrested three individuals linked with the twin bombings at the Gateway of India and Zaveri Bazaar in Mumbai in August of that year (Rediff.com 2009). Some of the largest scale arrests actually occur in Pakistan, after a major LeT operation results in international pressure, which prompts Pakistan to crack down. This occurred in the months after the 2008 Mumbai attacks, when the Pakistani government arrested hundreds of LeT members, although most were never charged and released months later (Tankel 2011b). Arrests also include the dozens of militants annually who voluntarily turn themselves into authorities in Jammu and Kashmir, some of whom enter state rehabilitation programs (The Hindu 2006).
- *Arrest Warrants*. This variable refers to when arrest warrants were issued against LeT personnel. Most, but not all, of the arrest warrants for LeT personnel are long-standing Indian arrest warrants for LeT leaders. The scope of these warrants expanded dramatically after the Mumbai attacks when the Indian government identified 34 figures linked to LeT that were sought in connection to the attacks (Indianexpress.com 2010). Under international pressure, Pakistan has also issued arrest warrants for LeT figures, as it did in its own dossier on the Mumbai attacks (IBNLive 2009). Arrest warrants have also been issued by the government of Jammu and Kashmir in the wake of major terrorist attacks, such as in September 2009 after a car-bomb detonated in Srinagar and police began a major operation to arrest the attack mastermind (Indianexpress.com 2009).
- *Asset Freezes*. Periodically, governments have taken actions that freeze assets of LeT and its affiliate organizations such as Jamaat ud-Dawa (JuD). Because LeT has no official presence in India, asset freezes are actually conducted by the Pakistani government—usually after a major LeT action that sparks international pressure. After the attack on India's parliament in December 2001 and the November 2008 Mumbai assault, the State Bank of Pakistan froze the accounts of LeT and its front groups (South Asia Terrorism Portal 2011a).
- *Kills*. Military or police operations occasionally kill LeT personnel. This variable includes LeT personnel killed in armed clashes, executed by the justice system, and killed by security forces. It does not include LeT personnel killed by

other non-state armed groups, criminals, armed civilians, or in intra-group conflict. Most of LeT's deaths described by this variable occur in Jammu and Kashmir where LeT has carried out extensive armed operations, although there are also LeT operatives killed by the government elsewhere in India.

• *Organizational Ban*. Long banned in India, LeT is also periodically banned by the Pakistani government (e.g., after the Mumbai attacks). The first occurrence of a Pakistani government ban of LeT occurred in January 2002 after the attack on India's parliament (and in the wake of 9/11) that nearly brought India and Pakistan to the brink of war. Shortly afterwards, LeT re-emerged as Jamaat ud-Dawa which continued operating LeT's social service networks (BBC News 2010). Pakistan has periodically reiterated these bans, and LeT in turn has organized new front operations.

• *Raids*. There are frequent government raids on suspected LeT hideouts and cells.

This section discusses the rules automatically found by the TP-Rule Extraction Engine that relate to attacks on Hindus and civilians and government actions against LeT.

In fact, the data mining system extracted the following TP-rule which shows that raids against LeT are often followed shortly thereafter by attacks on civilians.

TP-Rule AAH-3. LeT carried out 1–3 attacks against civilians one month after months in which:

• LeT claimed to be a religious organization and
• There were 0–12 raids on LeT.

Support = 10
Probability = 100 %,   *Inverse   Probability* = 100 %,   *Negative Probability* = 0 %

Note that TP-rule (AAH-3) above and TP-rule (AAH-4) below both make reference to LeT being a religious organization. Readers familiar with LeT and its brand of Ahl Hadith Islam are certain to recognize that LeT has been a religious organization for most of its existence.

TP-rule (AAH-4) below is very similar to (AAH-3) and applies to *arrests* of LeT personnel instead of raids against them.

TP-Rule AAH-4. LeT carried out 1–3 attacks against civilians one month after months in which:

• LeT claimed to be a religious organization and
• 0–6 LeT personnel were arrested.

Support = 10
Probability = 100 %,   Inverse   Probability = 100 %,   Negative Probability = 0 %

Though this TP-rule has the highest possible probability, inverse probability, and lowest possible negative probability, rule (AAH-5) below shows that a similar rule also applies to two-month offsets. This rule also provides further evidence for the hypothesis advanced above (Sect. 4.1) that there is a relationship between deaths of LeT commanders and subsequent attacks by LeT on Hindus and civilians.

TP-Rule AAH-5. LeT carried out 1–9 attacks against Hindus two months after months in which:

- LeT made territorial claims and
- 0–4 LeT personnel were arrested.

Support = 10
Probability = 90.9 %,   *Inverse*   *Probability* = 90.9 %,   *Negative*
Probability = 33.3 %

Similar rules are derived using the variable for arrest warrants against LeT personnel being issued.

TP-Rule AAH-6. LeT carried out 1–3 attacks against civilians two months after months in which:

- LeT's basic organization was not dissolved and
- Arrest warrants for 1–39 LeT personnel were issued.

Support = 11
Probability = 100 %,   *Inverse*   *Probability* = 100 %,   *Negative*
Probability = 0 %

A related TP-rule says that even if the number of arrest warrants issued is 1, the probability of LeT attacks on civilians 2 months later is still 100 %.

The TP-rules linking attacks on Hindus and arrests of LeT personnel are further strengthened by the following rule with a 3-year time offset.

TP-Rule AAH-7. LeT carried out 1–9 attacks against Hindus three months after months in which:

- LeT did not make territorial claims and
- Arrest warrants for 0–7 LeT personnel were issued.

Support = 10
Probability = 90.9 %,   *Inverse*   *Probability* = 90.9 %,   *Negative*
Probability = 33.3 %

TP-rules (AAH-4)-(AAH-7) provide considerable support for the hypothesis that arrests of LeT personnel in a given month are often followed (note: not "caused") within 1–3 months by LeT attacks on Hindus.

In the same way, when LeT personnel are killed by government action (rather than raids, arrests, or arrest warrants) there is considerable evidence to support the hypothesis that attacks against Hindus and civilians will occur in the next 1–3 months.

TP-Rule AAH-8. LeT carried out 1–3 attacks against civilians one month after months in which:

- LeT claimed to be a religious organization and.
- 0–10 LeT personnel were killed by the government.

Support = 10
Probability = 100 %,   *Inverse   Probability* = 100 %,   *Negative Probability* = 0 %

The same trend continues in looking at government bans of LeT.

TP-Rule AAH-9. LeT carried out 1–3 attacks against civilians one month after months in which:

- LeT claimed to be a religious organization and
- There was a (partial) government ban.

Support = 10
Probability = 100 %,   *Inverse   Probability* = 100 %,   *Negative Probability* = 0 %

Finally, there was a weaker relationship between asset *freezes* against LeT and their propensity to carry out attacks against Hindus and civilians.

This subsection ends by noting that there appears to be a very strong inverse relationship between government efforts to take legal action against LeT (e.g., via arrests, arrest warrants, bans on the organization, kills of LeT members—and more weakly, asset freezes) and LeT's propensity to carry out attacks on Hindus and civilians (primarily in Kashmir) in the following 1–3 months. This is not to claim that LeT exhibits this behavior intentionally—only that they behave this way— either intentionally or unintentionally.

4.3 Attacks Against Civilians and Hindus and a Defactionalized LeT

Our system also automatically derived TP-rules which strongly imply a strong link between LeT's ability to stay de-factionalized (no splitting, no splintering, etc.) and its attacks on civilians (without specific reference to religion that we could identify) as well as attacks on Hindus in particular.

TP-Rule AAH-10. LeT carried out 1–3 attacks against civilians one month after months in which:

- LeT claimed to be a religious organization and
- There was no splintering within LeT.

Support $= 10$
Probability $= 100\ \%$,   *Inverse   Probability* $= 100\ \%$,   *Negative Probability* $= 0\ \%$

Although terrorist organizations are often prone to internal strife and factionalization, LeT has been relatively successful at avoiding this fate. Another Pakistani jihadi group, Jaish-e-Mohammed (JeM) is practically a case study on splits and internal conflict. It was formed out of a split with another organization and underwent a series of internal splits. Nearly all of the Pakistani terrorist groups faced a challenge when the government of Pakistan aligned with the United States against al-Qaeda. These organizations were caught between their loyalty to their sponsors in the Pakistani military and their ideological affiliation with al-Qaeda and the Taliban (as well as their antipathy towards the United States). These splits led breakaway factions of JeM to target the Pakistani state (South Asia Terrorism Portal 2011b; Mir 2003)[5] (including an assassination attempt on then President Musharraf in 2003) (Imran 2004). Lashkar-e-Taiba has only had one formal (and brief) split and has refrained from turning its guns on the state. But, there have been numerous reports of lower-level splits and intra-group strains. Stephen Tankel argues that the competing priorities of loyalty to the state and sympathy for al-Qaeda and the Taliban have introduced tension into the organization that can lead to higher-profile attacks against "acceptable" targets (Tankel 2011b). Another important cause of factions can be personality driven, rather then ideological. In 2004, in the group's only formal split, LeT's co-founder Zafar Iqbal and his supporters left the organization for a time because he felt that Hafez Saeed was

---

[5] A description of JeM's founding from the invaluable South Asia Terrorism Portal can be found here—http://www.satp.org/satporgtp/countries/india/states/jandk/terrorist_outfits/jaish_e_mohammad_mujahideen_e_tanzeem.htm.

For background on JeM's splits in the years after 9/11 see—"The Maulana's Scattered Beads," Amir Mir, *OutlookIndia.com*, September 1, 2003—http://www.outlookindia.com/article.aspx?221267.

favoring his own clique in distributing key positions and distributing the organizations funding (Rana 2004). Iqbal has since returned to LeT. But periods of internal strife can also lead to reductions in organizational activity as the leadership is distracted by group politics. A third possibility is that Pakistan's Inter Services Intelligence (ISI) agency intentionally fosters intra-organizational conflict and splitting so as to "divide and rule" LeT and check Hafez Saeed's power base or, works with LeT to confuse authorities and evade international sanctions. Overall, LeT organizational stability may have helped it to continue to carry out its violent activities in the manner of its choosing.

TP-Rule AAH-11. LeT carried out 1–3 attacks against Hindus one month after months in which:

- LeT claimed to be a religious organization and
- LeT was not involved in any intra-organizational conflict.

Support = 10
Probability = 100 %,   *Inverse*   *Probability* = 100 %,   *Negative*
Probability = 0 %

Though the above rule is about attacks on civilians, we derived an almost identical TP-rule where the "There was no splintering within LeT" condition in the above TP-rule was replaced by "There was no inter-organizational conflict." In this case, we derived a mirror image of this TP-rule, applicable to attacks targeted specifically towards Hindus rather than just civilians.

A similar rule (about attacks on civilians) says that if LeT was not in the process of splintering and just one LeT member was arrested during a given month, we could expect attacks on civilians within 2 months.

TP-Rule AAH-12. LeT carried out 1–3 attacks against civilians two months after months in which

- LeT was not splitting/dissolving and
- Exactly one LeT member was arrested.

Support = 10
Probability = 100 %,   *Inverse*   *Probability* = 100 %,   *Negative*
Probability = 0 %

The hypothesis that LeT carries out attacks on Hindus 1–3 months after months in which it was not splitting is also supported when considering a 3-month offset as shown by the following rule.

> **TP-Rule AAH-13.** LeT carried out attacks against Hindus three months after months in which:
>
> • LeT was not splitting
>
> *Support* = 21.
> *Probability* = 95.5 %,   *Inverse*   *Probability* = 100 %,   *Negative*
> *Probability* = 0 %

This rule has very strong support and high probability, inverse probability, and negative probability.

When LeT is not splitting or enjoys a period of cohesive unity without any intra-organizational conflict, they tend to carry out attacks on civilians—mostly Hindus. For much of the period during which the Kashmir conflict was particularly active, attacks on civilians were a regular occurrence. As the fighting in Kashmir has cooled, these attacks have decreased. However, organizational splits may be a tool by which the ISI restrains LeT which would explain why these events correlate.

4.4 Attacks Against Hindus and Desertion by LeT Members

A rather unusual set of TP-rules generated by the system dealt with the case of desertions by LeT members. Since 2002, on over a dozen occasions, LeT operatives in Jammu & Kashmir have turned themselves over to security forces rather then carry out their attacks (South Asia Terrorism Portal 2011a).[6] This is particularly remarkable because LeT indoctrination emphasizes the glory of martyrdom in fedayeen operations. These kinds of attacks are LeT trademarks in which the attacker intends to fight to the death (but will escape if practical). These operations are distinct from suicide bombings in that the attacker will not die by his own hand. This distinction has been the subject of extensive discourse among radical Islamist leaders (Rana 2006).[7] Given the depth of LeT training and the number of operatives who do in fact fight to the death, the willingness of operatives to surrender may be a significant indicator.

---

[6] Several instances can be found in SATP's chronology—www.satp.org.

[7] (Rana 2006, pp. 337–339) excerpts a 2001 essay by a top LeT leader on this issue.

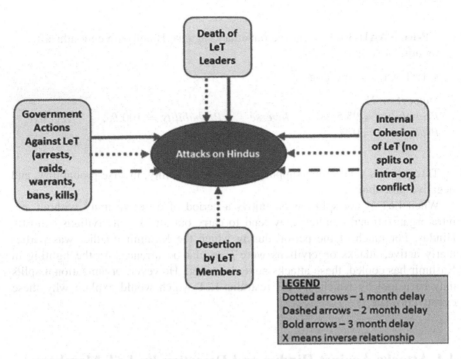

Fig. 4.1 Summary of variables related to LeT attacks on Hindus

TP-Rule AAH-14. LeT carries out one attack against Hindus one month after months in which:

- No LeT commanders died and
- Between 1 and 3 LeT members deserted.

Support = 15.
Probability = 100 %,   *Inverse*   *Probability* = 100 %,   *Negative*
Probability = 0 %

A related TP-rule derived by the system says this result holds even if up to 6 LeT commanders died, rather than none.

4.5 Conclusion and Policy Options

This chapter concludes with a summary of the main findings. Figure 4.1 above shows a succinct summary of the findings, specifying the variables that seem to be related to LeT attacks on civilians and those which target individuals on the basis of religious belief—usually targeting at Hindus, but also on some occasions Sikhs and Jews.

Whether looking 1, 2, or 3 months into the future, there is strong evidence that deaths of LeT commanders lead to subsequent attacks on Hindus (or unspecified civilians) in the following 1–3 months;

When LeT is internally cohesive and no splintering/splitting/intra-organizational conflict occurs, then this is a leading indicator (by at least 1–3 months) of attacks on civilians;

When LeT is under pressure from the government via government actions such as arrests, issuance of arrest warrants, asset freezes, bans on the organization, or explicit kills of LeT operatives, it reacts with violence against Hindus within a 1–3 month period. It is not clear that government counter-measures cause these attacks, but it does suggest that these standard counterterror strategies, used by governments worldwide, have limited utility against LeT. This will be discussed further in Chap. 11 on policies that may stand a chance of reining in LeT.

Last but not least, when LeT members desert (even in small numbers), there seems to be a good chance of LeT attacks on Hindus in subsequent months.

It is tempting to conclude from these results that in order to reduce LeT or LeT-backed attacks on civilians, one could try to:

Increase governmental pressure on LeT by more rigorous law enforcement and police/military actions such as government bans on LeT, issuance of arrest warrants for—and arrests of—LeT members, kills of LeT members and cadres;

Increase targeted attacks on LeT commanders with the hope of eliminating them;

Increase the use of methods to cause splintering within LeT and to promote intra-organizational rivalry within LeT;

Increase methods to support desertion by LeT members—such a strategy has been used very successfully by Rwandan President Paul Kagame to disarm the Hutu militias which swarmed into the Democratic Republic of Congo, and the result has been a huge success in reducing the ability of the Forces Democratiques de Liberation du Rwanda (FDLR—or Democratic Forces for the Liberation of Rwanda).

However, this chapter has only examined one class of actions carried out by LeT—namely attacks on civilians in general and Hindus specifically. Use of these methods may have other adverse consequences—and when formulating a strategy to deal with LeT, one must globally consider all possible outcomes of the proposed strategy or strategies. This approach will be taken at the end of this book in Chap. 10.

References

Ahmad, M. (2002). *Militants kill 9 Amarnath pilgrims, injure 32 Rediff.com.* August 6, 2002. Retrieved from http://www.rediff.com/news/2002/aug/06jk.htm

BBC News (2010). *Profile: Lashkar-e-Taiba.* May 3, 2010, Retrieved from http://news.bbc.co.uk/2/hi/south_asia/3181925.stm

IBNLive (2009). *Pak dossier pins down LeT men for 26/11.* July 18, 2009. Retrieved from http://ibnlive.in.com/news/its-official-pak-dossier-pins-down-let-men-for-2611/97429-2.html

Imran, M. (2004). *Jaish splinter planned Musharraf assassination*. The Daily Times. January 23, 2004. Retrieved from http://www.dailytimes.com.pk/default.asp?page=story_23-1-2004_pg7_3

Indianexpress.com (2009). *LeT, Hizbul teamed up for J-K car blast*. September 14, 2009. Retrieved from http://www.indianexpress.com/news/let-hizbul-teamed-up-for-jk-car-blast/516722/

Indianexpress.com (2010). *Indo-Pak talks: India priorities terror, Saeed's arrest*. February 25, 2010. Retrieved from http://www.indianexpress.com/news/indopak-talks-india-prioritises-terror-saeeds-arrest/584333/0

Mir, A. (2003) *The Maulana's Scattered Beads, Amir Mir, OutlookIndia.com*. September 1, 2003. Retrieved from http://www.outlookindia.com/article.aspx?221267

National Counterterrorism Center's World Wide Incidents Tracking System. http://wits.nctc.gov

Rana, A. (2004). *Jamaatud Dawa splits*. The Daily Times, July 18, 2004. Retrieved from http://www.dailytimes.com.pk/default.asp?page=story_18-7-2004_pg7_20

Rana, MA. (2006). *A to Z of Jihadi Organizations in Pakistan*. Translated by Saba Ansan. Pakistan: Mashal Books. Retrieved from http://www.desistore.com/jehadiorg.html

Rediff.com (2009). *A chronology of the 2003 Mumbai twin blasts case*. July 27, 2009. Retrieved from http://news.rediff.com/report/2009/jul/27/a-chronology-of-the-2003-mumbai-twin-blasts-case.htm

South Asia Terrorism Portal (2011a). *Incidents involving LeT. Institute for Conflict Management New Delhi*. http://www.satp.org/satporgtp/countries/india/states/jandk/terrorist_outfits/lashkar_e_toiba_lt.htm

South Asia Terrorism Portal (2011b). *Jaish-e-Mohammed (Army of the Prophet). Institute for Conflict Management, New Delhi*. http://www.satp.org/satporgtp/countries/india/states/jandk/terrorist_outfits/jaish_e_mohammad_mujahideen_e_tanzeem.htm

Tankel, S. (2011a) *Lashkar-e-Taiba: Past operations and future prospects*. National security studies program policy paper. Washington: New America Foundation, April 2011.

Tankel, S. (2011b). *Storming the World Stage: The Story of Lashkar-e-Taiba*. London: C. Hurst & Co.

The Hindu (2006). *More J&K militants surrender*. December 14, 2006. Retrieved from http://www.jammu-kashmir.com/archives/archives2006/kashmir20061214b.html

United Press International (2002). *Hindus massacre shrouds Indian talks offer*. January 1, 2002. Retrieved from http://www.upi.com/Business_News/Security-Industry/2002/01/01/Hindus-massacre-shrouds-Indian-talks-offer/UPI-57141009894624/

Chapter 5
Attacks Against Public Sites, Tourist Sites and Transportation Facilities

Abstract This chapter describes the conditions under which LeT carries out terrorist attacks against three types of targets: public sites, tourist sites, and transportation facilities such as railway stations and airports. The chapter discusses several TP-rules about these types of terror attacks that were derived automatically from the LeT data set used in this book.

LeT has carried out numerous spectacular attacks against public sites, tourism sites, and transportation facilities and networks. Besides the carnage of these attacks, the targets are often carefully selected because they are symbols of India's growing power and prosperity. Examples of such attacks include:

- LeT carried out and claimed responsibility for the Deember 23, 2000 attack on the Red Fort (or *Lal Qila*) in Delhi, which killed 3 Indian security personnel. In 2005 an Indian court found 7 individuals guilty in the attack including one Pakistani national (*BBC News* 2005). The Red Fort in Delhi is a magnificent red sandstone structure with extensive walls and stunning buildings within it. Built by the Mughal emperor Shah Jahan (who also built the Taj Mahal), the Red Fort is one of Delhi's major tourist attractions. It is also a symbol of the Indian government. Every August 15 (the date of India's independence from British colonial rule), India's Prime Minister addresses the nation from the ramparts of the Red Fort. That the palace built by a great Muslim emperor at the height of Mughal power is now associated with the prime minister of Hindu India fuels the rage of radical Islamists such as LeT. Raising the flag of Islam in Delhi and over the Red Fort in particular had long been a jihadi ambition (Abou Zahab 2007).[1]
- LeT has made numerous attacks on transportation hubs in Jammu and Kashmir. In January 2001, a fedayeen attack on Srinagar Airport killed four security personnel, two civilians, and six LeT members. August of that year saw attacks

---

[1] Because the Indian Army administered the Red Fort until 2003, the LeT claimed that its attack the Red Fort was on a military, rather than a civilian target.

V. S. Subrahmanian et al., *Computational Analysis of Terrorist Groups: Lashkar-e-Taiba*,   95
DOI: 10.1007/978-1-4614-4769-6_5, © Springer Science+Business Media New York 2013

against the Jammu Railway Station (in which 12 were killed including a LeT operative) and a thwarted attack on the airport in Jammu. There have also been innumerable attacks on bus stands throughout Jammu and Kashmir (South Asia Terrorism Portal 2011).

- On July 11, 2006 a series of seven bombs were detonated on Mumbai's densely packed commuter train system, killing about 200 people dead and injuring over 1000 people. The Indian government held LeT responsible for this attack. (Daily News & Analysis 2011). While there is some dispute about LeT's role in the attack, the other leading suspect is the Indian Islamist organization Indian Mujahideen (IM), which receives support from LeT.
- The infamous November 26, 2008 Mumbai attacks targeted public infrastructure as well as tourism-related sites. The Chattrapati Shivaji train station—affectionately known as VT (or Victoria Terminus) to Mumbai's natives—was the first site struck by the LeT attackers, followed subsequently by attacks at the iconic Taj Mahal hotel in Colaba, and the Oberoi and Trident hotels in Nariman Point, the less well-known Chabad House, as well as several other less iconic sites. The Taj Mahal, Oberoi, and Trident hotels are world-class hotels catering to international businessmen and Western tourists that symbolize Mumbai's rise as an international financial center.

This chapter examines relevant TP-rules in an effort to identify the environmental conditions under which LeT carries out attacks such as those mentioned above on public sites, tourist sites, and transportation networks.

5.1 Attacks on Public, Symbolic and Tourist Sites

The system derived several TP-rules associated with attacks on symbolic and tourist sites. First, all TP-rules derived here only applied to 3 month offsets, not to 1 or 2 month offsets. Many of the attacks covered in these rules required extensive long-term planning. According to David Headley's testimony in the *US v. Rana*, Headley spent several years reconnoitering targets for the November 2008 assault on Mumbai (Rotella 2011). Rules that examine the longer 3 month offset timeframe may help indicate specific triggers to a large-scale attack.

TP-Rule PST-1. There are 0–29 deaths in LeT attacks against symbolic sites three month after months in which:

- LeT was not splitting.

Support = 100
Probability = 100 %,   *Inverse*   *Probability* = 100 %,   *Negative*
Probability = 0 %

LeT has carried out several high-profile attacks against highly symbolic sites such as the December 2000 attack on the Red Fort, the September 2002 attack on the temple at Akshardham and the July 2005 attack on a makeshift temple at Ayodhya. In some cases the casualties from these attacks are relatively low, in the 2002 attack on Ayodhya the only casualties were the LeT attackers who were all killed by security forces before they could attack the worshippers (Rediff.com 2005). Other attacks include the Akshardham attack in which LeT attackers did gain entry to the temple and killed 29 worshippers (Joseph 2002). Regardless of the casualties, these attacks are significant; they represent LeT efforts to take the conflict beyond Jammu and Kashmir both geographically and psychologically. By attacking symbols of the state of India or the Hindu religion, LeT is attempting to actualize its rhetoric calling for the destruction of the Indian state and the subjugation of the Hindu religion.

These attacks are rare (there have been less than a dozen months in which these kinds of attacks occur), thus the lower bound of casualties (0) in many cases represents months in which there was no attack on a symbolic target (although some attacks, such as the attempted Ayodhya Temple assault, did not cause any casualties.) However, they do not occur in the periods after LeT has split (or experienced substantial intra-organizational conflict).

There are two possible explanations for this behavior on the part of LeT. First, splits, the result of strife within the organization, distract the organization's leadership and hamper planning. A second possibility is that the splits are often efforts to create new front organizations in order to evade international scrutiny. In December 2001, after the attack on India's parliament (and given the high-level of scrutiny on terrorist organizations after 9/11) the armed components of LeT were supposedly split from the charitable components. A more complicated split occurred in July 2004, Zafar Iqbal, LeT's co-founder, reportedly left the organization and established Khairun Naas ("the most excellent of people"). There were conflicting reports on this incident. Iqbal had been frustrated with Hafez Saeed's leadership for several years. Iqbal felt that Saeed had favored relatives and members of his caste for promotion, had taken a much younger second wife, and was not distributing the organization's funds fairly. According to some sources, Iqbal had the support of key leaders, including leaders from LeT's jihadi wing such as Zaki-ur-Rehman Lakhvi and was laying to claim to LeT's headquarters at Muridke. But other sources from within LeT said that either the government engineered the split to reduce LeT activities or LeT itself was adopting a new identity to reduce pressure (Rana 2004). A few years later Iqbal rejoined LeT. Compared to other Pakistani terrorist groups LeT has been stable. In the words of C. Christine Fair, "...LeT has never experienced a leadership split of any consequence since its founding. While it has at various times reorganized, this is not the same thing as dividing into opposing factions because of leadership quarrels (Fair 2011)." However, there have been anecdotal reports saying that splits within LeT were orchestrated by the Pakistani Inter Services Intelligence agency in order to ensure that LeT stays firmly within ISI control and to place a check on Hafez Saeed's power base.

Whether or not the decline in high-profile attacks after organizational splits is the result of internal disorder or external pressure, it still represents an important policy option for reducing LeT violent activity.

In addition to the above rules that describe conditions under which LeT carried out attacks on public sites, symbolic sites, and tourist sites, we also derived conditions when LeT did not carry out such attacks. These rules are useful as they suggest policies that might enable policy-makers to reduce attacks by LeT on such sites.

The first TP-rule below says that LeT does not carry out attacks against such targets when 0–5 of LeT's commanders died 3 months back. This rule suggests that targeting top LeT operational commanders may in fact lead to a loss of leadership within the organization, hampering the ability to execute complex attacks with only low-level foot soldiers.

TP-Rule PST-2. LeT carries out 0 attacks on symbolic/tourist sites three months after months in which:

- LeT had locations across the border from India and
- 0–5 leaders of LeT died.

Support = 40.
Probability = 90.9 %,   *Inverse   Probability* = 100 %,   *Negative Probability* = 0 %.

This rule has very strong support—the condition identified was known to be true in 40 months and in 90.9 % of those months, LeT did not attack symbolic/tourist sites, although it should be noted that attacks on symbolic and tourist targets are relatively rare. As mentioned above, a probable explanation might be that the loss of leadership temporarily left the organization "re-grouping" and "re-organizing" so as to re-coup those capabilities. This rule is further strengthened by another TP-rule we derived with a 2-month offset.

Though the next rule does not have as strong statistics as TP-Rule (PST-2), it adds corroborating evidence that when LeT has locations across the border from India and when 0–4 LeT commanders died, this is followed by a few months that exhibit a down-turn in the number of attacks on symbolic/tourist sites within India.

TP-Rule PST-3. LeT carries out 0 attacks on symbolic/tourist sites two months after months in which:

- LeT had locations across the border from India and.
- 0–4 leaders of LeT died.

Support = 40.
Probability = 97.6 %,   *Inverse   Probability* = 95.2 %,   *Negative Probability* = 40 %.

The next rule indicates that when the government released 0–9 LeT members, these attacks dropped down to 0 after 3 months.

TP-Rule PST-4. LeT carries out 0 attacks on symbolic/tourist sites three months after months in which:

- LeT had locations across the border from India and
- 0–9 LeT members were released by the government.

Support = 40
Probability = 90.9 %,   *Inverse   Probability* = 100 %,   *Negative Probability* = 0 %

The release of arrested LeT operatives may be helpful (at least temporarily) in reducing attacks on public/tourist sites, possibly suggesting the existence of a *quid pro quo* in which LeT promises not to carry out attacks on public/tourist sites "in exchange" for release of some of its people. However, many of the arrest releases are cases in which LeT leader Hafez Mohammad Saeed is released from house arrest or imprisonment. Often he was initially arrested after a major attack on a target in India, but later released (Hindustan Times 2009). Major attacks on targets within India are resource intensive for LeT and bring international pressure to bear on the Pakistani government. Thus, for months and even years after a major attack, LeT may lie low and refrain from large-scale attacks.

TP-Rule PST-5. LeT carries out 0 attacks on symbolic/tourist sites three months after months in which:

- LeT had locations across the border from India and
- 0–15 members of LeT were killed by the government.

Support = 40
Probability = 90.9 %,   *Inverse   Probability* = 100 %,   *Negative Probability* = 0 %

The next rule shows that when LeT has locations across the border from India and 0–15 LeT members are killed, then there is a high probability that LeT will not attack symbolic/tourist sites in India in 3 months.

This is yet another powerful rule that significantly connects ongoing government operations against LeT with a reduction in attacks by LeT on symbolic or public sites. When large numbers of LeT operatives are killed, it is usually due to clashes (discussed in Chap. 9) with Indian security forces in Jammu and Kashmir. It may indicate a reduced capability, but it may also indicate an LeT focus on operations in Jammu and Kashmir.

5.2 Attacks on Transportation Facilities

As mentioned above, LeT has carried out numerous attacks against transportation facilities including the Srinagar airport attack in January 2001 and the attack on Mumbai's Chattrapati Shivaji Terminus on November 26, 2008. The system was not able to derive a large number of rules connecting attacks on transportation facilities/networks with the prevailing environment in which LeT operates.

Nonetheless, the system generated a pair of very interesting rules that describe when LeT was not carrying out attacks on transportation facilities—and both of these rules are connected to government operations against LeT.

The first rule below connects deaths of LeT commanders with "no attacks" on transportation facilities.

TP-Rule PST-6. LeT carries out 0 attacks on transportation facilities/networks three months after months in which:

- 0–6 commanders of LeT died.

Support = 106
Probability = 86.9 %,   *Inverse   Probability* = 100 %,   *Negative Probability* = 0 %

This rule has robust support, but a slightly lower probability than some of the rules previously derived. Nevertheless, it strongly implies that deaths of LeT commanders (often occurring in armed clashes with security forces) are a positive factor in reducing attacks on transportation facilities/networks. Deaths of LeT commanders in clashes are often indicative of effective government efforts at finding and neutralizing LeT units. After losing several commanders, LeT may not have the resources to launch offensive operations.

TP-Rule PST-7. LeT carries out 0 attacks on transportation facilities/networks three months after months in which:

- 0–6 commanders of LeT died
- 0–14 LeT operatives were killed by the government.

Support = 104
Probability = 87.4 %,   *Inverse   Probability* = 98.1 %,   *Negative Probability* = 0 %

A related rule says that when 0–6 LeT commanders died in a given month and the government killed 0–14 LeT operatives, the probability of 0 attacks on transportation facilities is very high.

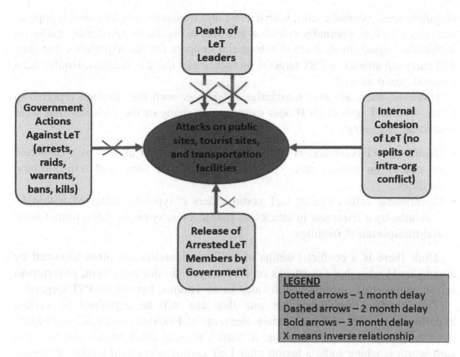

Fig. 5.1 Summary of variables related to LeT attacks on public sites, tourist sites, and transportation facilities

This TP-rule has statistics that are similar to the preceding rule—the two rules jointly suggest that government action against LeT has the desired impact of reducing attacks against transportation facilities 3 months hence.

5.3   Conclusion and Policy Options

Figure 5.1 above shows the relationship between LeT attacks on symbolic sites, tourist and tourism-related sites, and transportation facilities on the one hand—and key variables that appear related to these attacks on the other hand. Figure 5.1 below shows the relationship between LeT attacks on symbolic sites, tourist and tourism-related sites, and transportation facilities on the one hand—and key variables that appear related to these attacks on the other hand.

In considering the variables that seem highly related to LeT attacks on public sites, symbolic sites, tourist sites, and transportation facilities, it appears that when LeT is internally cohesive and has no splits or splintering going on within the organization, there is substantial support for the hypothesis that they will carry out attacks on PST targets—in fact, under these conditions, Hindus are a targeted group as well.In considering the variables that seem highly related to LeT attacks on public sites, symbolic sites, tourist sites, and transportation facilities, it appears that when LeT is internally cohesive and has no splits or splintering going on within the organization, there is substantial support for the hypothesis that they will carry out attacks on PST targets—in fact, under these conditions, Hindus are a targeted group as well.

However, there are also substantial *differences* with the situation reported in Chap. 4 on LeT attacks on Hindus even though many of the variables that seem related are the same:

- Deaths of LeT commanders are usually followed in 1–3 months by a *reduction* in attacks on public sites, symbolic sites, tourist sites, and transportation facilities;
- Government action against LeT commanders is typically followed within 1–3 months by a *reduction* in attacks on public sites, symbolic sites, tourist sites, and transportation facilities.

Thus, there is a conflict—deaths of LeT commanders are often followed by attacks on Hindus, but not attacks on PST targets. In the same vein, government action against LeT are followed by attacks on Hindus, but not on PST targets.

This yields a *dichotomy*, but one that can still be explained by certain hypotheses. Perhaps LeT and certain elements of Pakistan's Inter Services Intelligence agency feel that Hindus are a "soft", easy to attack target, and that it is well worth teaching India a lesson after LeT commanders died and/or after government actions against LeT and LeT personnel were enacted.

After considering the results of Chaps. 4 and 5, it appears that the strategy of increasing internal dissension within LeT's ranks is a good way of destabilizing LeT, at least in so far as reducing attacks on Hindus and reducing attacks on public sites, symbolic sites, tourist sites, and transportation facilities are concerned.

Whether this is a good "comprehensive" strategy remains to be seen and will be explored as we continue with this book.

References

Abou Zahab, M. (2007). 'I shall be waiting for you at the door of paradise:' The Pakistani Martyrs of the Lashkar-e Taiba. In Aparna Rao, Michael Bollig, & Monika Bock (Eds.), *The Practice of War: Production, Reproduction and Communication of Armed Violence* (pp. 117–126). New York: Berghahn Books.

BBC News (2005). *Red Fort attack—seven convicted.* October 24, 2005. Retrieved from http://news.bbc.co.uk/2/hi/south_asia/4370866.stm

Daily News & Analysis (2011). *Mumbai train blast: legal battle to end soon.* July 10, 2011. Retrieved from http://www.dnaindia.com/india/report_mumbai-train-blast-legal-battle-to-end-soon_1564296

Christine, FC. (2011). Lashkar-e-Tayiba and the Pakistani State. *Survival, 53*(4), 29–52.

Hindustan Times (2009). *Hafez Saeed's Profile.* June 2, 2009. Retrieved from http://www.hindustantimes.com/india/Hafiz-Saeed-s-Profile/548305/H1-Article1-417093.aspx

Joseph, J. (2002). *Terrorists had entered temple, ready for long haul, Rediff.com.* September 26, 2002. Retrieved from http://www.rediff.com/news/2002/sep/25guj22.htm

Rana, A. (2004). *Jamaatud Dawa splits, Daily Times.* July 18, 2004. Retrieved from http://www.dailytimes.com.pk/default.asp?page=story_18-7-2004_pg7_20

Rediff.com (2005). *6 militants storm Ayodhya, killed.* July 5, 2005. Retrieved from http://in.rediff.com/news/2005/jul/05ayo.htm

Rotella, S. (2011). *The American Behind India's 9/11-And How U.S. Botched Chances to Stop Him.* Retrieved from http://www.propublica.org/article/david-headley-homegrown-terrorist

South Asia Terrorism Portal (2011). *Incidents involving LeT, Institute for Conflict Management New Delhi.* Retrieved from http://www.satp.org/satporgtp/countries/india/states/jandk/terrorist_outfits/lashkar_e_toiba_lt.htm

Chapter 6
Attacks Against Professional Security Forces

Abstract LeT has carried out numerous attacks against professional security forces—primarily the Indian Army and police. This chapter describes the conditions under which LeT carries out terrorist attacks against professional security forces. The chapter discusses several TP-rules about these types of terror attacks that were derived automatically from the LeT data set used in this book.

Over the years, LeT has carried out numerous attacks against professional security forces—a term we define broadly to include attacks against local police forces, against national security forces such as India's Army, against paramilitary national government forces such as India's Border Security Force (BSF) and Central Reserve Police Force (CRPF). These are attacks that specifically target security forces and include various tactics such as ambushes, sniper and grenade attacks, and IED attacks targeting security forces. The key to this variable is that the initiative is with LeT and that security forces were specifically targeted. Some of these attacks include:

- In September 2009, Jammu and Kashmir police accused LeT of setting off a car bomb in Srinagar (the capital of Jammu and Kashmir) that took the lives of two police officers and a bystander. Security forces were obviously the targets as the bomb was set off when a police bus passed (Thaindian News 2009).
- A month before, LeT operatives, working with Jaish-e-Mohammed and Hizb-ul-Mujahedeen members, carried out a pair of attacks in Srinagar. In the first attack two Central Reserve Police Force (CPRF) troops were shot and killed at close range with pistols fitted with silencers. In the second attack, a kilometer away, a grenade was hurled at a CPRF vehicle (Pandit 2009).
- In May 2002 terrorists opened fire on a bus and then stormed the Kalu Chak army camp, firing automatic weapons and lobbing hand-grenades. Ultimately, 33 people were killed including women and children (the camp was home to officers' families) (Ahmad 2002). The terrorist group al-Mansoorian, which is generally believed to be a front for LeT, took responsibility for the attack (Tribune News Service 2002).

- In March 2001 a joint LeT, JeM and HM team ambushed security forces in Rajouri in Jammu and Kashmir, killing fifteen (South Asia Terrorism Portal 2011).

  Our principal findings in this chapter indicate that:

- LeT carries out attacks on professional security forces when they have a strong relationship with the Pakistani military (this should come as no surprise);
- LeT carries out attacks on professional security forces within 2 months after periods when they are publishing periodicals to get their message out and when none of their commanders were killed;
- LeT carries out attacks on professional security forces within 2 months after periods when they are publishing to get their message out and when there is no strong aggressive government action against them such as arrests/bans and other instruments of force;
- A subtle variant of the above rules says that LeT carries out attacks against professional security forces 3 months after months in which one of their commanders died and in which the government arrested 1–2 members.

6.1 Attacks on Professional Security Forces and Relationship with Pakistani Military

There is extensive evidence supporting the hypothesis that the Pakistani military (including the Inter Services Intelligence agency) use terror groups such as the Haqqani network, Jaish-e-Mohammed and Lashkar-e-Taiba (Tankel 2011)[1] as proxies in their campaigns against their perceived adversaries. In the case of LeT, this has contributed to a large number of attacks by LeT against (primarily) Indian security forces.

> **TP-Rule PSF-1**. LeT carries out 1 attack against professional security forces 2 months after months in which:
>
> - LeT received military support from the Pakistani government.
>
> *Support* = 14
> *Probability* = 87.5 %, *Inverse Probability* = 100 %,
> *Negative Probability* = 0 %

---

[1] (Tankel 2011, pp. 60–61) describes ISI support for LeT, and how it made LeT combat proficient. On p. 64, Tankel cites contacts who state that the ISI helped establish JeM as a counter-balance to LeT. On p. 99, he discusses how the ISI acted to coordinate LeT and the Haqqani network.

The first TP-rule says that there is a strong relationship between attacks on Indian security forces and the links LeT has with the Pakistani military. This rule has relatively strong support—the pre-condition was true for 14 months. Of these 14 months, there was an 87.5 % probability that two months after there was documentary evidence asserting a strong relationship between LeT and the Pakistani military, there would be attacks on professional security forces. The inverse probability was 100 % and the negative probability was 0 %, indicating a rule with strong statistics.

Anyone familiar with the last 20 years of Pakistani politics and strategy toward India could argue that it is "common knowledge" that Pakistani military support has emboldened LeT. However, this argument is not necessarily correct. Chapter 5 does not find any TP-rule that asserts that LeT is emboldened to attack tourist sites or transportation facilities when there is evidence of increased support from the Pakistani military—the reason is that the data does not support it. Instead, it supports the hypothesis that when there is specific documentation of a relationship between LeT and the Pakistani military, the attacks that took place 2 months later were primarily on Indian professional security forces, not against Hindus or against tourist sites/transportation facilities. While there is little question that training by the Pakistani military has made LeT a more capable organization overall, it appears that military support (which includes tactical support in infiltrating the Jammu and Kashmir border) does not necessarily extend to the large-scale LeT operations in India (the 2008 Mumbai assault in which ISI handlers trained LeT's attack team may be an exception). This would fit with the Pakistani military's overall strategy of keeping the conflict with India simmering, but without letting it boil over into large-scale war. Thus the drivers for the large-scale attacks may not be supported by Pakistani policy, but the result of some other factors. Whatever the causes, financiers and trainers of terrorist groups must be held responsible for the terrorist acts carried out by the organizations they support.

The system also derived another related rule which said that when there was no government ban and LeT had a strong relationship with the Pakistani military, then there was strong support for the hypothesis that LeT would attack professional security forces 2 months later.

TP-Rule PSF-2. LeT carries out 1 attack against professional security forces 2 months after months in which:

- There was no government ban on LeT and
- LeT had a strong relationship with the Pakistani military.

Support = 13
Probability = 92.9 %, *Inverse Probability* = 92.9 %,
Negative Probability = 33.3 %

This TP-rule has slightly less support than the preceding TP-rule but it has a higher probability, while at the same time, also having a higher negative probability.

The government bans tracked here are those by the Pakistani government, which formally bans LeT under international pressure (as occurred after the 2001 attack on the Indian parliament). After the 2008 Mumbai attacks, JuD (LeT's main front group) was subjected to a substantial, albeit brief crackdown, as well as sanctioned by international bodies, but not formally banned by the Pakistani government (Tankel 2011). However, LeT usually reemerges soon after under a new name, although with little question about the organization's real identity. This rule suggests that when LeT is not under political pressure from the Pakistani government—which in turn is being pressured by third parties like the US—LeT and their ISI sponsors are emboldened to increase attacks against Indian security forces, thereby keeping the situation in Jammu and Kashmir unstable.

6.2 Attacks on Professional Security Forces and Use of Communications Media and Government Action Against LeT

The system found considerable evidence for the hypothesis that LeT attacks professional security forces 2–3 months after months when two things occur concurrently:

- LeT is waging a media campaign (e.g., through the news media and through the periodicals and magazines that they publish);
- LeT is facing government aggression.

TP-Rule PSF-3. LeT carries out 1 attack against professional security forces 3 months after months in which:

- LeT was waging a communications campaign using news/periodicals to get their message out

Support = 14
Probability = 87.5%, *Inverse Probability* = 93.3%,
Negative Probability = 33.3 %

Our first rule specifies that when LeT is waging a media campaign using news/ periodicals, then this means that there is a high probability that they will carry out attacks on professional security forces 3 months later.

TP-Rule (PSF-3) is just one example of a TP-rule that shows a relationship between when LeT carries out attacks against professional security forces and LeT's publicity campaigns. LeT runs a large media arm that publishes a number of journals and books. Their flagship publication *Mujalla-ul-Dawa* is estimated to have weekly circulation of 80,000. Fitting with LeT's overall worldview of both waging jihad and converting people to their version of Islam, the publications include articles about making Islam relevant to people's lives *and* extensive descriptions of LeT's armed operations, particularly in Kashmir (Rana 2006, pp. 327–332).[2] These descriptions focus on LeT operations against military targets. It is interesting that publishing efforts appear to expand in periods before new rounds of military operations. They could be laying the groundwork among LeT supporters for further violent campaigns.

Another related rule specifies a similar relationship between LeT's waging a communication campaign when 0–2 LeT members died.

TP-Rule PSF-4. LeT carries out 1 attack against professional security forces 2 months after months in which:

- LeT was waging a communications campaign using news/periodicals to get their message out and
- 0–2 LeT operatives were arrested.

Support = 17
Probability = 89.5 %, *Inverse Probability* = 68.0 %,
Negative Probability = 47.1 %

Together with TP-Rule (PSF-3), this rule provides even further support for the hypothesis that there is a strong relationship between LeT waging a communications campaign using news and periodicals and attacks on Indian professional security forces 2–3 months later.

A closely related TP-rule that we derived further supplements this rule, though in the case when no LeT commanders died.

TP-Rule (PSF-5) has better "statistics" than (PSF-4) and adds further support to the hypothesis that LeT carries out armed attacks against professional security forces shortly after they embark on a publicity campaign using news/periodicals. In fact, all of these rules jointly suggest a relationship between waging a communications campaign and attacks on professional security forces. This rule is also intriguing in that it indicates that periods when LeT's battlefield commanders are not killed are periods that see attacks on security forces, reinforcing the hypothesis that eliminating these commanders reduces LeT's capabilities.

---

[2] See also (Abou Zahab 2007) who did her research using accounts of martyrdom in LeT publications.

TP-Rule PSF-5. LeT carries out 1 attack against professional security forces 2 months after months in which:

- LeT was waging a communications campaign using news/periodicals to get their message out and
- 0 LeT leaders died.

Support = 23
Probability = 88.5 %, *Inverse Probability* = 92.0 %,
Negative Probability = 20 %

The system also derived two additional rules that establish a relationship between LeT attacks on professional security forces and situations where LeT was carrying out a communications campaign and LeT commanders were killed by government security forces 3 months earlier.

TP-Rule PSF-6. LeT carries out 1 attack against professional security forces 3 months after months in which:

- LeT was waging a communications campaign using news/periodicals to get their message out and
- 0–5 LeT members were killed by government action.

Support = 14
Probability = 93.3 %,   *Inverse*   *Probability* = 93.3 %,   *Negative*
Probability = 25 %

This TP-rule implies a very strong relationship between months in which (i) LeT was carrying out a communications campaign and (ii) government action killed 0–5 LeT members and the propensity of LeT to attack professional (Indian) security forces 3 months later. The deaths of LeT members, as opposed to commanders, may reflect higher numbers of infiltrations and thus greater resources for attacking security forces.

TP-Rule PSF-7. LeT carries out 1 attack against professional security forces three months after months in which:

• LeT was waging a communications campaign using news/periodicals at level 1 to get their message out and
• 0–5 LeT members were killed by government action.

Support = 15
Probability = 88.2 %,   *Inverse*   *Probability* = 100 %,   *Negative*
Probability = 0 %

6.3 Attacks on Professional Security Forces and Situations When LeT's Leadership is Functionally Differentiated

One of the variables tracked about LeT pertained to how LeT's leadership structure is organized. LeT's organization is "functionally differentiated" when there is a clear separation of roles/duties for the leadership. For instance, during the past few years, the role of Hafez Mohammed Saeed is to preach fiery sermons supportive of LeT's mission, while the role of Zaki-ur-Rehman Lakhvi is believed to be focused on planning military operations, while other figures manage LeT's fundraising and social service operations.

It is something of a surprise that this variable—taken in conjunction with variables related to the release of LeT members previously arrested—seems to presage attacks by LeT on Indian security forces.

TP-Rule PSF-8. LeT carries out attacks against professional security forces 3 months after months in which:

• LeT's leadership was functionally differentiated and
• 1–2 arrested LeT members were released by the government

Support = 12
Probability = 92.3 %, *Inverse Probability* = 75 %,
Negative Probability = 44.4 %

This TP-rule says that LeT carries out attacks against professional security forces 3 months after months in which their leadership clearly defined and separated the roles of its leaders and when 1–2 LeT members who had previously been arrested were released. As mentioned above the figure most often released is Hafez Saeed. This rule indicates that while Saeed is imprisoned, the organization—due to the functional differentiation of its leadership—is able to continue functioning, although it may reduce operations under governmental pressure. When that pressure is reduced and Hafez Saeed is released, LeT can begin preparing another round of violence.

The system derived another similar rule when exactly one LeT member died. The statistics reported in TP-Rule (PSF-8) remained completely unchanged.

6.4 Attacks on Professional Security Forces and Desertions by LeT Members

We derived TP-rules specifying various conditions under which LeT does **not** attack professional Indian security forces. First and foremost, the system automatically derived rules specifying that LeT does not carry out such attacks 3 months after desertion by one or more members.

> **TP-Rule PSF-9**. LeT carries out 0 attacks against professional security forces 3 months after months in which:
>
> • 1 LeT member deserted.
> *Support* = 11
> *Probability* = 91.7 %,   *Inverse   Probability* = 91.7 %,   *Negative Probability* = 25 %

This provides some evidence for the hypothesis advanced already in Chap. 5 that the use of methods to encourage desertion by LeT members may place pressures on LeT that hamper its operational ability.

6.5 Attacks on Professional Security Forces and Issuance of Arrest Warrants for LeT Members

We also derived TP-rules further supporting the hypothesis that government action against LeT members causes a reduction in attacks on professional security forces a month later.

> **TP-Rule PSF-10**. LeT carries out 0 attacks against local security installations 1 month after months in which:
>
> • Arrest warrants for 1–39 members of LeT were issued and
> • No splintering was going on.
>
> *Support* = 99
> *Probability* = 90 %,   *Inverse   Probability* = 100 %,   *Negative Probability* = 0 %

In fact, we derived a multiplicity of such rules (and similar related rules).

TP-Rule PSF-11. LeT carries out 0–1 attacks against professional national security forces 1 month after months in which:

- 0–16 members of LeT were killed by the government and
- No splintering was going on.

Support = 29
Probability = 87.9 %,   *Inverse*   *Probability* = 100 %,   *Negative*
Probability = 0 %

Virtually the same rule applies when the condition that the government killed 0–16 members of LeT is replaced by the condition that the government arrested 0–24 LeT members.

6.6 Conclusion and Policy Options

Figure 6.1 below shows the relationship between the individual variables studied in this chapter and LeT's propensity to attack professional security forces, usually those of India.

The results of this chapter allow us to conclude that:

- The *Relationship with the Pakistani military* is critical to LeT's attempts to carry out attacks on India security forces—though this is not unexpected;
- *Major communications campaigns* by LeT appear to be correlated with future attacks by LeT on professional security forces;
- *Release of arrested LeT members by the government* seems unhelpful in reducing the number of attacks by LeT on professional security forces (however, we saw earlier in Chap. 5 that the release of LeT members seems connected to a reduction in the number of attacks by LeT on tourist sites, symbolic sites, and transportation facilities);
- *Functional differentiation and clean separation of duties between LeT's leadership* appears to increase the probability of attacks on Indian professional security forces;
- *Desertion* by LeT members seems to help (as it did in Chap. 5) in reducing the number of attacks by LeT on professional security forces;
- *Government action against LeT including bans, arrest warrants, arrests, and kills of LeT members* are of mixed value when trying to reduce attacks by LeT against professional security forces.

As seen previously in Chap. 5, this suggests several policies against LeT. However, a policy that helps to reduce the number of attacks on professional

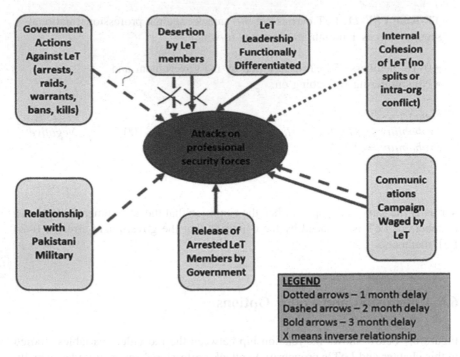

Fig. 6.1 Summary of variables related to LeT attacks on professional security forces

security forces may merely cause LeT to resort to some other type of attack or target—therefore comprehensive policy suggestions, together with an analysis of the pros and cons of those policy options, will wait until Chap. 10.

- *Encouraging Desertion.* As in the case after the Hutu-orchestrated in Rwanda, the ability to "incentivize" LeT cadres to defect has the potential to destabilize the organization and reduce its ability to carry out attacks, especially if other actions can be brought to bear concurrently.
- *Covert action against LeT media campaigns.* LeT's media campaigns are closely related to attacks on Indian security forces—any covert action that can help disrupt these campaigns would help prevent LeT from making claims of "legitimacy" that enable them to find financial support, recruits, and operational support for some of their actions.
- *Disrupting functional differentiation amongst LeT leadership.* LeT's leaders are able to execute carefully planned attacks on Indian professional security forces when they are able to split up their leadership duties neatly amongst the leaders. Disrupting such a well-oiled and neatly functioning machine by disabling some of the operatives may pay dividends, especially as there is some evidence that under certain circumstances (though not all), government action against LeT leaders may prove counter-productive. Another option is to combine the strategy

of encouraging desertion and disrupting the LeT leadership by weaning away (through suitable incentives), key LeT leaders. Such strategies could include providing amnesty from prosecution for past acts in exchange for an agreement to be monitored electronically and provide intelligence about their group. A widely publicized strategy of this sort, even with only a few successes, also has the potential to sow "distrust" between LeT leaders and cadres, leading to a potential disruption of the ability of the organization to work harmoniously.

There are some precedents for eliminating terrorist groups by exacerbating their internal tensions and playing them off against other groups. Most famously, the Abu Nidal Organization—perhaps the world's most notorious terrorist group in the 1980s—was effectively de-fanged by intelligence operations that persuaded Abu Nidal that his top aides were betraying him. Crucial to this operation was intelligence from another terrorist organization, the PLO, which had a bloody rivalry with Abu Nidal (Perry 1992).

References

Abou Zahab, M. (2007). I shall be waiting for you at the door of paradise:' The Pakistani Martyrs of the Lashkar-e Taiba, in edited by R. Aparna, M. Bollig, M. Bock (Eds.) *The Practice of War: Production, Reproduction and Communication of Armed Violence*, (pp. 117–126) New York: Berghahn Books

Ahmad, M. (2002). 33 killed in attack on army camp in Jammu. *Rediff.com*. Retrieved May 14, 2002, from http://www.rediff.com/news/2002/may/14jk.htm

Pandit, M. S. (2009). Terrorists strike near J&K assembly, 2 jawans killed. *The Times of India*. Retrieved September 1, 2009, from http://articles.timesofindia.indiatimes.com/2009-09-01/india/28093464_1_crpf-jawan-crpf-personnel-crpf-vehicle

Perry, M. (1992). *Eclipse: The last days of the CIA*. New York: Morrow.

Rana, M. A. (2006). *A to Z of Jihadi Organizations in Pakistan*. Translated by Saba Ansan. Pakistan: Mashal Books http://www.desistore.com/jehadiorg.html

South Asia Terrorism Portal (2011). Incidents involving LeT. Institute for Conflict Management New Delhi. http://www.satp.org/satporgtp/countries/india/states/jandk/terrorist_outfits/lashkar _e_toiba_lt.htm

Tankel, S. (2011). *Storming the world stage: The story of Lashkar-e-Taiba*. London: C. Hurst & Co.

Thaindian News (2009). Three killed in Srinagar blast, police blame LeT. http://www.thaindian.com/newsportal/india-news/three-killed-in-srinagar-blast-police-blame-let-fourth-lead_100246647.html

Tribune News Service (2002). Al-Mansoorian behind attack. Retrieved October 1, 2002, from http://www.tribuneindia.com/2002/20021002/main2.htm

Chapter 7
Attacks Against Security Installations and Infrastructure

Abstract This chapter describes the conditions under which LeT carries out terrorist attacks against security installations such as military bases, police stations, and checkpoints. The chapter discusses several TP-rules about these types of terror attacks that were derived automatically from the LeT data set used in this book.

Over the years, LeT has carried out numerous attacks against security installations and security infrastructure. These attacks include attacks against numerous police stations, bases or facilities belonging to security organizations such as the Indian Army, the CRPF, or Border Security Force. Attacks on security personnel, which were discussed in the previous chapter, were often hit and run attacks on security patrols or guards. Attacks on installations were frequently fedayeen (suicide attackers) attacks in which a small party of LeT attackers opened fire on a security facility or attempted to enter the facility and kill those within. A few examples follow.

- On November 3, 1999, LeT terrorists attacked the headquarters of the Army's 15 Corps in Badami Bagh near Kashmir's capital in Srinagar, killing eight soldiers on the Corps Public Relations staff (along with the two terrorists) (Swami 1999).
- On March 3, 2001 a pair of LeT operatives struck an Army camp at Baramulla in Jammu and Kashmir. Firing and throwing grenades, the two attackers killed four soldiers before being killed (The Asian Age 2001).
- In December 2004, a pair of LeT fedayeen attempted to enter a Border Security Force encampment at Sopore, but were killed before they could do so (Chakravarti 2005).
- In July 2007, two LeT operatives attempted to enter a CRPF battalion headquarters at the defunct Bhabha Atomic Research facility (this is different from the Bhabha Atomic Research Center facility in Trombay outside Mumbai) on the outskirts of Srinagar. The LeT operatives were killed before they could enter the encampment, although they injured several security personnel (The Hindu 2007).

V. S. Subrahmanian et al., *Computational Analysis of Terrorist Groups: Lashkar-e-Taiba*,   117
DOI: 10.1007/978-1-4614-4769-6_7, © Springer Science+Business Media New York 2013

- These attacks are not limited to Jammu & Kashmir. The Indian government accused LeT of perpetrating a January 2008 attack on a CRPF recruitment center in Rampur in Uttar Pradesh that left seven CRPF personnel and a civilian dead (The Times of India 2008).

These represent a small sample of attacks carried out by Lashkar-e-Taiba against security installations—mostly, but not exclusively, in Jammu and Kashmir.

The rest of this chapter studies the circumstances under which LeT carries out attacks against security installations. The principal findings are listed below:

- As in the case of attacks against professional security forces, attacks against (primarily Indian) security infrastructure typically occur after periods when LeT has been conducting a media campaign using news/periodicals;
- Smaller scale arrests of LeT personnel (less then 20 in a month) are typically followed by attacks against Indian security infrastructure a couple of months later. In the same way, lack of government action against LeT such as raids and asset freezes seem to lead to attacks against Indian security infrastructure;
- When none of LeT commanders died and when there are no Pakistani government bans against LeT, one can expect attacks against Indian security installations within a couple of months;
- Periods when LeT was providing political support to other Islamist groups around the world coincided with attacks on Indian security installations;
- Last but not least, LeT attacks Indian security installations within a couple of months of periods when there is internal cohesion (i.e., no intra-organizational conflict) within LeT.

7.1 Attacks Against Security Infrastructure and Publicity Campaigns by LeT

There is strong evidence that LeT carries out attacks against Indian security infrastructure within 2–3 months of periods during which LeT is ramping up publicity efforts through its network of publications.

> **TP-Rule SI-1.** LeT carries out attacks against Indian security installations two months after months in which:
>
> - LeT waged a publicity campaign using news and periodicals to get their message out and
> - None of LeT's leaders were arrested.
>
> *Support* = 12
> *Probability* = 92.3 %,   *Inverse   Probability* = 100 %,   *Negative Probability* = 0 %

This rule suggests that in periods when LeT's leadership is intact (and thus operational planning capabilities, as well as a clear chain of command exists), and when LeT waged a publicity campaign using news and periodicals to drum up support for their positions, attacks on Indian security installations two months later are likely. One possible conclusion is that attacks on Indian security installations are "standard operating procedure" for LeT. Arrests of LeT leadership are often situations when Pakistan detains LeT's leader Hafez Mohammed Saeed. During these periods, which usually follow a major LeT attack in India, LeT is pressured to reduce its operations. LeT also has an extensive publications wing. Its articles regularly feature tales of LeT's armed exploits in Jammu and Kashmir. When LeT magazines are being published and the leaders are free to plan, then LeT carries out attacks on Indian security installations.

Another TP-rule we derived along similar lines says that LeT attacks Indian security infrastructure when they have launched a media campaign and when no splintering is going on within the group.

TP-Rule SI-2. LeT carries out 1 attack against Indian security installations two months after months in which:

- LeT waged a publicity campaign using news and periodicals to get their message out and
- LeT was not splintering.

Support = 10
Probability = 100 %,   *Inverse*   *Probability* = 83.3 %,   *Negative*
Probability = 28.6 %

As in the case of the preceding rule, this rule suggests that when nothing is interfering with LeT operations, then it is free to launch attacks on Indian security. LeT has only had limited cases of splitting due to internal factions. These splits, as well as struggles for controlling the organization's assets, can reduce the effectiveness of the organization's ability to carry out attacks—particularly strikes against relatively hard targets like security installations.

TP-Rule SI-3. LeT carries out attacks against Indian security installations three months after months in which:

- LeT waged a publicity campaign using news and periodicals to get their message out and
- No LeT members were arrested.

Support = 14
Probability = 87.5 %,   *Inverse*   *Probability* = 77.8 %,   *Negative*
Probability = 21.1 %

A third TP-rule we derived looks at the link between LeT operations and *no arrests* being made of LeT personnel.

Again, this TP-rule provides support for the hypothesis that when LeT is operating without either internal or external pressures, and when they have launched a media campaign, there is a high likelihood that they will carry out attacks against Indian security installations within three months.

7.2 Attacks on Security Infrastructure, Publicity Campaigns by LeT, and Lack of Action by the Government Against LeT

This section continues the discussion started in the preceding section, but focuses on the situations when:

- LeT is waging a publicity campaign and
- The government has not taken aggressive action against LeT.

When there is no government ban against LeT (and thus LeT is allowed to function with impunity within Pakistan), and when there is a strong publicity campaign being orchestrated by LeT via the news media and periodicals, there is a high probability of attacks on Indian security installations within two months.

This powerful rule clearly shows that when LeT is not under threat from the Pakistani government and when they have ratcheted up rhetoric in print media and periodicals, attacks against Indian security installations will probably follow within 3 months.

> **TP-Rule SI-4.** LeT carries out attacks against Indian security installations two months after months in which:
>
> - LeT waged a publicity campaign using news and periodicals to get their message out
> - The Pakistani government does not ban LeT.
>
> *Support* $= 12$
> *Probability* $= 92.3\ \%$,   *Inverse*   *Probability* $= 100\ \%$,   *Negative*
> *Probability* $= 0\ \%$

When there is no government ban and when LeT is leading a communications campaign, the probability of LeT attacks on Indian security installations is very high.

TP-Rule (SI-4) suggests that strong anti-LeT action from the Pakistani government could be helpful in reducing LeT attacks against Indian security installations. However, this is rare as the Pakistani military, which has effectively controlled Pakistan for the last few decades, values LeT as a proxy and has been unwilling to forcefully crack-down on its activities. The frequency of LeT's print media campaigns as a variable in LeT attacks suggests an alternative strategy might be to disrupt or otherwise hamper LeT's ability to express its views in the print media and via its extensive publishing network.

In addition to situations when the Pakistani government does not ban LeT, there are similar patterns when Pakistani government inaction against LeT (and at times an effective carte-blanche or even tacit encouragement to move forward with attacks) provides the group with the ability to orchestrate attacks against Indian security installations. At the same time, a "motive" for such attacks is orchestrated by LeT waging an effective public relations campaign via the print media and periodicals. Pakistani government "inaction" refers to periods when there are a very small number of raids and/or when very few LeT personnel are arrested.

> **TP-Rule SI-5.** LeT carries out attacks against Indian security installations two months after months in which:
>
> - LeT waged a publicity campaign using news and periodicals to get their message out and
> - 0–19 LeT members are arrested.
>
> *Support* = 12
> *Probability* = 92.3 %, *Inverse* *Probability* = 100 %, *Negative* *Probability* = 0 %

This TP-rule shows that when the number of LeT operatives arrested is small and when LeT is waging a publicity campaign using print media and periodicals, their motive (through the message being disseminated through print media) and means (through unfettered ability to carry out their attacks due to government inaction) to carry out attacks is high. It is important to note that most months when there are a relatively small number of arrests, those arrests are carried out by Indian security. In this TP-rule, the range of arrests is under 20, excluding the months in which Pakistan undertook large-scale crackdowns in which dozens, or even hundreds, of LeT operatives were arrested. These rules indicate that Pakistani measures against LeT *can* be effective at reducing LeT operations, when that government chooses to implement them.

A similar rule applies in the case when there are no asset freezes and when LeT is conducting a publicity campaign. When the Pakistani government does not freeze LeT's assets, it has the ability to finance attacks using freely flowing funds to pay for personnel, materiel, and transportation costs. This gives it the "means" to carry out the attacks, while the "motive" is provided in the publicity campaign being waged. While LeT has been under consistent financial sanctions by the

United States Department of Treasury, the organization has been skilled at evading the impact of these sanctions (U.S. Dept. of Treasury 2010). LeT has used front organizations and changed its financial management personnel when US sanctions have interfered with their ability to operate. Locally enforced sanctions, by a committed regulator with deep knowledge of LeT operations on the ground might prove more effective.[1]

TP-Rule SI-6. LeT carries out attacks against Indian security installations two months after months in which:

- LeT waged a publicity campaign uses news and periodicals to get their message out and
- No assets of LeT or its leaders were frozen by the Pakistani government.

Support = 11
Probability = 100 %,　　　*Inverse*　　　*Probability* = 91.7 %,　　　*Negative*
Probability = 16.7 %

TP-Rules (SI-1)-(SI-6) jointly provide compelling evidence that strong government action is needed to rein LeT in—and moreover, that strong action is needed to prevent LeT from using print media and periodicals to get their message out. In a separate paper that uses a separate set of computational techniques based on game theory (Dickerson et al. 2011), some of the authors have suggested that to rein in LeT, two things need to happen concurrently:

- Increased covert action by the US and
- Either increased covert action by India or increased coercive diplomacy by India.

The results of this section suggest that covert action could target LeT's access to the print media and/or periodicals, as well as increasing targeting LeT operatives in order to disrupt the ability of the organization to plan terrorist attacks.

7.3 Attacks on Security Infrastructure and Political Support from Pakistan

The system derived TP-rules that show strong technical support for the hypothesis that LeT's propensity to carry out attacks against Indian security installations is

---

[1] Finding such a committed regulator who has both the knowledge and the ability to study LeT's financial operations will be difficult. An alternative approach might be to penetrate LeT's IT systems via covert action and then use sophisticated financial analysts to study the data unearthed.

closely tied to periods when LeT was actively and vocally supporting jihadi groups within Pakistan and worldwide.

TP-Rule SI-7. LeT carries out 1-3 attacks against Indian security installations two months after months in which:

- LeT was providing political support to other Islamist organizations
- LeT was not engaged in any intra-organizational conflict.

Support = 10
Probability = 90.9 %,   *Inverse*   *Probability* = 100 %,   *Negative*
Probability = 0 %

This is just one of several rules the system derived. Other similar rules replaced the "no intra-organizational conflict" condition by a similar condition requiring that "no splintering" was occurring.

Another similar rule is shown below.

TP-Rule SI-8. LeT carries out 1-3 attacks against Indian security installations two months after months in which:

- LeT received political support from the Pakistani government and
- No LeT leaders resigned.

Support = 10
Probability = 90.9 %,   *Inverse*   *Probability* = 100 %,   *Negative*
Probability = 0 %

The system derived another rule virtually identical to the above rule, which replaces the "no LeT leaders resigned" by the condition "Between 0 and 1 LeT commanders died".

This suite of TP-rules focuses on the years when LeT was actively and vocally engaged with broader pan-Islamist movements (including al-Qaeda). LeT provided support for Islamists in Chechnya, Bosnia, and across the Middle East. This support, which came to a peak at the turn of the millennium and continued until a few months after 9/11, coincided with some of the heaviest fighting by LeT in Jammu and Kashmir. After 9/11 it became less prudent for LeT to engage publicly with other radical Islamist organizations and its political support waned. During this period, Islamists felt that historical trends favored their cause and LeT was inspired to press forward with its violent activities in Jammu and Kashmir, as well as extend them into the rest of India.

7.4 Conclusion and Policy Options

Figure 7.1 below summarizes the relationship between LeT's attacks on Indian security installations and several variables.

This figure shows us that:

- *LeT's use of the media (*via *print media and periodicals)* is a precursor of LeT attacks on Indian security installations which typically tend to occur within a couple of months. LeT frequently extols fedayeen attacks in its publications. The February 2001 issue of LeT's flagship magazine *Mujjala-ul-Dawa* contained an extensive description of the December 2000 attack on the Red Fort in Delhi (because it was administered by the army and had barracks on location, the Red Fort was coded both as a tourist site and a security installation in our data set). In the May 2001 issue of *Mujjala-ul-Dawa*, a senior LeT leader discussed the attack on the army headquarters at Badami Bagh (among others) (Rana 2006).[2] There were attacks on security installations in the months following these publications. As in the case of Chap. 6, where attacks on professional security forces were often mounted within 2–3 months of a publicity campaign waged by LeT, here too we typically see a 2-month delay. The articles both justified LeT's operations in terms of Islamic law and Pakistan's honor, as well as glorified the operations. In this way the publicity campaigns are recruiting tools and they keep the LeT's operations fresh in the minds of its supporters.
- *LeT is emboldened to attack Indian security installations when operations against them are largely shut down.* Rules (SI-1), (SI-2), (SI-3), (SI-4)-(SI-6) and (SI-8) all point to situations where there are no bans on LeT, no asset freezes, no deaths of LeT commanders, and no resignations by LeT leaders. With the exception of TP-Rule (SI-5), all of these rules are unambiguous and support this hypothesis. TP-Rule (SI-5) supports the possibility that such attacks go on even when as many as 19 LeT personnel are arrested and thus introduces some contrary evidence to the hypothesis that LeT consistently attacks security installations two months after periods when LeT was not actively suppressed. However, the maximum range of arrests is 19, and in many cases is much fewer, so this may not indicate a period when Pakistan's government was exerting itself to prevent LeT operations.
- *LeT carries out attacks on Indian security installations within 2 months of periods when it was providing political support to other Islamist groups.* There is clear evidence that periods when LeT is vocally active in international Islamist circles are followed within approximately 2 months by attacks on Indian security installations. This rule does dovetail with the previous rules. When LeT is free to operate so openly that it can support other organizations

---

[2] Both examples are taken from (Rana, pages 330–339).

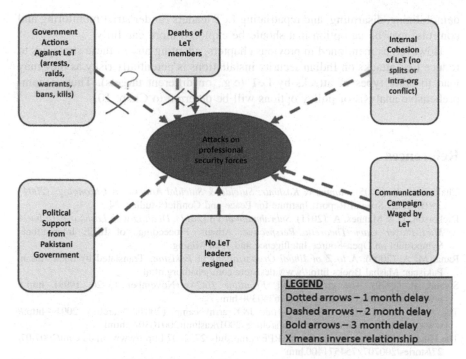

Fig. 7.1 Summary of variables related to LeT attacks on security installations and infrastructure

(some outside of Pakistan) it is also free to plot attacks against hard targets such as security installations.

- *Internal cohesion of LeT as an organization* is also clearly tied to its propensity to carry out attacks against Indian security installations. Some rules such as (SI-2) clearly indicate that when LeT is not splintering or engaged in intra-organizational conflict, it is more likely to attack Indian security installations. In addition, TP-Rule (SI-8) which applies to LeT attacks on security installations during periods when there were no resignations by LeT leadership suggests also that there were no significant intra-organizational squabbles 2 months prior to times when such attacks were launched by LeT.

As in the case of Chap. 6, these TP-rules offer a small set of possible policy options against LeT.

Clearly, disrupting LeT's public media outreach ability has the potential to help disrupt LeT's predilection to target Indian security installations by denying them the ability to espouse their views and grievances prior to an attack and later use those grievances to justify the attack.

Helping foment intra-organizational conflict or additional splintering within LeT also has the potential to disrupt LeT operations against Indian security installations. A strategy suggested earlier in Chap. 4 that attempts to "wean away" LeT leaders and cadres, possibly by following a Rwanda-style strategy of demobilizing, disarming, and repatriating LeT leaders (under strict monitoring and controls) may be an option that should be explored more carefully.

However, as mentioned in previous chapters, adopting any of these strategies to reduce LeT attacks on Indian security installations is potentially risky as they may lead to other types of attacks by LeT (e.g., on different targets). Thus, a comprehensive analysis of policy options will be deferred to Chap. 10.

References

Chakravarti, R. (2005). *Jammu & Kashmir: Suicide & Suicidal Attacks—A Chronology (2004–2005)*. IPCS Special Report: Institute for Peace and Conflict Studies. 24.

Dickerson, J., & Mannes, A. (2011). *Subrahmanian VS (2011). Dealing with Lashkar-e-Taiba: A Multi-Player Game-Theoretic Perspective*. Athens: Proceedings of IEEE International Symposium on Open-Source Intelligence and Web Mining.

Rana, M. A. (2006). *A to Z of Jihadi Organizations in Pakistan*. Translated by Saba Ansan. Pakistan: Mashal Books http://www.desistore.com/jehadiorg.html

Swami, P. (1999). The growing toll, *Frontline 16*(24) (November 13–26, 1999). http://www.hinduonnet.com/fline/fl1624/16240390.htm

The Asian Age, 6 killed as militants raid J&K army camp. (2001). March 4, 2001—http://www.jammu-kashmir.com/archives/archives2001/kashmir20010304a.html

The Hindu. (2007). Attempt to storm CRPF camp, July 27, 2007 http://www.hindu.com/2007/07/27/stories/2007072754571400.htm

The Times of India, India Times News Network & Agencies. (2008). LeT behind militants attack on CRPF camp in UP, January 1, 2008—http://articles.timesofindia.indiatimes.com/2008-01-01/india/27761816_1_attack-on-crpf-camp-crpf-personnel-recruitment-camp

U.S. Dept. of Treasury. (2010). Treasury targets financial network of Pakistan-Based Terrorist Organization Lashkar-e-Tayyiba, Press Center, November 24, 2010—http://www.treasury.gov/press-center/press-releases/Pages/tg980.aspx

Chapter 8
Other Types of Attacks

Abstract This chapter describes the conditions under which LeT carries out terrorist attacks on holidays as well as attempted (but unsuccessful) attacks. The chapter discusses several TP-rules about these types of terror attacks that were derived automatically from the LeT data set used in this book.

This chapter studies assorted attacks carried out by LeT including attacks on holidays (such as India's Independence Day or religious festivals such as Diwali) and attempted attacks that were thwarted by security forces, usually Indian.

We derived a total of 19 TP-rules describing conditions under which LeT carried out **attacks on holidays**. Specifically, these rules were closely related to the following variables:

- *Lack of Arrests of LeT Leaders.* In periods when no LeT leaders were arrested and certain other conditions listed below were true, there was a good chance that these periods would be followed a few months later by attacks carried out on holidays.
- *Low levels of Aggressive Government Action Against LeT.* We found several rules that related variables associated with low levels of government action against LeT with attacks on holidays carried out 2 months later by LeT. This class of variables is also closely related to the previous class of variables about deaths of LeT commanders (obviously, when there is no ongoing government action against LeT, there was a lower probability of deaths of LeT commanders in those attacks).
- *Pressing for India to leave Jammu & Kashmir.* When LeT was forcefully calling for India to leave the province of Jammu & Kashmir, in conjunction with lack of government action against LeT, there was a high probability that LeT would carry out attacks on holidays.
- *Advocating Change of Lifestyle.* When LeT leaders advocated a change in lifestyle involving adopting a more strict interpretation of Islam in accordance with Ahl Hadith teachings and there is lack of concerted government action

V. S. Subrahmanian et al., *Computational Analysis of Terrorist Groups: Lashkar-e-Taiba*, 127
DOI: 10.1007/978-1-4614-4769-6_8, © Springer Science+Business Media New York 2013

against LeT, there is a higher probability that LeT will carry out attacks on holidays.

Finally, the system discovered four TP-rules that described conditions under which LeT attempted to execute attacks but were unsuccessful in doing so. The variables related to such attempted attacks include many we have seen so far:

- *Lack of resignations of LeT leaders* seems to be closely related to LeT's attempts to execute attacks within 1–2 months. LeT leaders usually resign as part of an effort to rebrand the organization after a major attack has brought international pressure on Pakistan. Thus these resignations coincide with government crackdowns and a break from attacks. Alternately, long periods when LeT leaders do not resign coincide with high levels of LeT activity.
- *Government Action Against LeT.* Government action against LeT including arrests and raids are also related to attempted attacks 1–3 months later.

In the rest of this chapter, we describe and study these rules in greater detail.

8.1 Attacks on Holidays by LeT

Launching attacks on major public holidays is a strategy that LeT has deployed on many occasions. Attacks on holidays can have a major symbolic value, it can cause general dread and terror, and striking at holiday crowds is a way to maximize casualties. One regular target of LeT attacks is the Republic Day celebrations in Jammu and Kashmir. Republic Day is celebrated on January 26 to commemorate the establishment of India's constitution—the celebrations in Delhi include a large parade down Rajpath, Delhi's most impressive street. The parade typically lasts a few hours and is a magnificent spectacle including displays of military hardware as well as a variety of cultural "floats". However, Republic Day celebrations occur across the country.

Thus, attacks on this day have been an obvious target for LeT which seeks to delegitimize Indian rule in Kashmir and ultimately undermine the Indian state itself. LeT has made numerous attempts to disrupt Republic Day in Jammu and Kashmir. January 2007 saw plots to cause disturbances on Republic Day in both Jammu and Kashmir and also in India's capital Delhi (*Daily News & Analysis* 2007). However, LeT attacks are not limited to secular Indian holidays. LeT has also struck Hindu and even Muslim festivals, such as the bombings on October 29, 2005, which struck two markets in Delhi and killed over 50. This bombing occurred the week before both the Hindu festival of Diwali and the Muslim festival of Eid, ensuring that the marketplaces were full of holiday shoppers (*BBC News* 2005).

The first TP-rule below is one of many similar TP-rules derived by the system relating to attacks by LeT on holidays following months when LeT advocates expelling India from Jammu and Kashmir and when LeT's leadership remains largely intact.

TP-Rule AOH-1. LeT carries out 1 attack on a holiday two months after months in which:

- LeT sought to remove Indian influence from Kashmir
- None of LeT's leaders arrested.

Support = 11
Probability = 91.7 %,   *Inverse   Probability* = 100 %,   *Negative Probability* = 0 %

The system derived several other TP-rules of a similar nature. For example, the TP-rule below replaces the condition "None of LeT's leaders were arrested" with "0–8 LeT members were killed by the government".

TP-Rule AOH-2. LeT carries out 1 attack on a holiday two months after months in which:

- LeT sought to remove Indian influence from Kashmir
- 0–8 LeT members were killed by the (Indian) government.

Support = 11
Probability = 91.7 %,   *Inverse   Probability* = 100 %,   *Negative Probability* = 0 %

To complete the picture of the relationship between attacks on holidays by LeT and the lack of offensive actions against LeT by the Indian (and Pakistani) governments, we now present one (of several) TP-rules linking the failure to arrest LeT leaders along with relatively few arrests of LeT personnel overall to attacks on holidays two months later.

TP-Rule AOH-3. LeT carries out 1 attack on a holiday two months after months in which:

- 0–8 LeT members were arrested and
- None of LeT's leaders were arrested.

Support = 10
Probability = 90.9 %,   *Inverse   Probability* = 90.9 %,   *Negative Probability* = 33.3 %

These three TP-rules provide considerable support to the hypothesis that LeT attacks on holidays occur two months after situations in which LeT is allowed to operate freely, with relatively few arrests of its members and when its leaders are permitted operate unhindered.

The smattering of TP-rules below show that the same variables have an impact when LeT is advocating a change of lifestyle as when it is pressing to expel India from Jammu and Kashmir. For LeT the missions of *dawa* (preaching) and *jihad* (holy war) are equal, essential, and reinforce one another. Through jihad, LeT seeks to liberate Muslim lands, and through dawa it seeks to bring Muslims to, what LeT views as, the proper interpretation of Islam. By focusing its efforts on the Kashmir conflict, which resonates among the Pakistani people, LeT's effort at jihad helps bring more Pakistanis to its beliefs. Missionary work among the Pakistanis, while primarily focused on propogating LeT's vision of Islam, also raises LeT's profile and can help in recruiting cadres to wage jihad (Tankel 2011). Thus advocating for a change in lifestyle in Pakistan and for the expulsion of India from Jammu & Kashmir goes hand in hand for LeT. Doing so without interference from authorities allows LeT to orchestrate major attacks on India.

TP-Rule AOH-4. LeT carries out 1 attack on a holiday two months after months in which:

- LeT was advocating a change in lifestyle and
- None of LeT's leaders arrested.

Support = 11
Probability = 91.7 %,   *Inverse*   *Probability* = 100 %,   *Negative*
Probability = 0 %

Adding further credibility to the hypothesis that LeT carries out such attacks when it is not the target of government action is the following TP-rule.

TP-Rule AOH-5. LeT carries out 1 attack on a holiday two months after months in which:

- LeT was advocating a change in lifestyle and
- 0–8 LeT personnel were killed.

Support = 11
Probability = 91.7 %,   *Inverse*   *Probability* = 100 %,   *Negative*
Probability = 0 %

TP-Rules (AOH-1) through (AOH-5) jointly imply that LeT attacks on holidays are closely related to LeT's being free to advocate both for the expulsion of India from Jammu and Kashmir and for Pakistanis to change their lifestyles to the Ahl

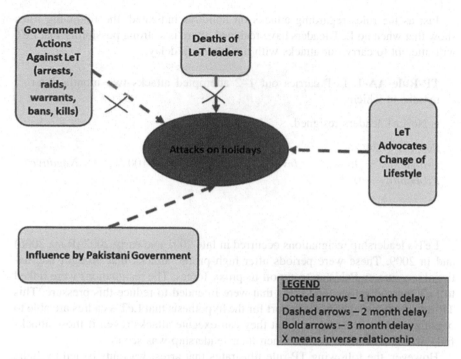

Fig. 8.1 Summary of variables related to LeT attacks on holidays

Hadith interpretation of Islam, along with a lack of government action against the group.

Figure 8.1 summarizes the relationship between LeT attacks on holidays and the variables discussed above.

8.2 Attempted Attacks by LeT

The final portion of this chapter deals with *attempted attacks* by LeT. This variable was coded when there was an attempt to execute an attack, but the attempt was unsuccessful. There have been a large number of high-profile LeT attacks that were intercepted before they reached their target including an August 2001 incident in which a pair of armed LeT operatives were killed near Jammu airport, on their way to carrying out a fedayeen attack. In another example, on June 28, 2007 three LeT terrorists were killed as they prepared to launch a suicide strike on an Army unit (South Asia Terrorism Portal 2011). Attempted attacks also occur elsewhere in India. In July 2005 a half-dozen heavily armed LeT operatives attempted to attack the temple at Ayodhya, but were killed before they could enter the complex (*Rediff.com* 2005). The fact that these attacks may have been unsuccessful does not detract from the fact that the attacks were attempted with every intention on LeT's part of causing grievous harm.

Just as the rules regarding attacks on holidays indicated, the following rules show that when no LeT leaders have resigned, there is a strong possibility that LeT will attempt to carry out attacks with a one month delay.

TP-Rule AA-1. LeT carries out 1–2 attempted attacks two months after months in which:

- No LeT leaders resigned.

Support = 24
Probability = 96 %,   *Inverse*   *Probability* = 100 %,   *Negative*
Probability = 0 %

LeT's leadership resignations occurred in late 2001 and early 2002 (Rana 2006) and in 2009. These were periods after high-profile attacks that brought international pressure on Pakistan to cut off its proxy forces. The resignations were linked to LeT reorganizations and splits that were intended to reduce this pressure. This TP-rule provides additional support for the hypothesis that LeT's cadres are able to organize and have confidence that they can execute attacks (even if those attacks fail) two months after months when their leadership was secure.

However, the following TP-rule illustrates that arrest warrants issued by India are not a deterrent to LeT's efforts to launch deadly attacks.

TP-Rule AA-2. LeT carries out 1–2 attempted attacks three months after months in which:

- Arrest warrants are issued for 1–34 of its members.

Support = 21
Probability = 95 %,   *Inverse*   *Probability* = 100 %,   *Negative*
Probability = 0 %

This rule is further strengthened by TP-Rule (AA-3) below, which states that a small number of government raids are not effective in stopping LeT's attempts to carry out attacks. Raids are situations in which the government has the initiative and includes attacks on LeT bases, safehouses, and the discovery of arms caches. Nonetheless, it appears that the LeT network is robust enough to continue operating after a limited crackdown.

TP-Rule AA-3. LeT carries out 1–2 attempted attacks one month after months in which:

- 0–2 raids against LeT are carried out by the government.

Support = 24
Probability = 96 %,   *Inverse*   *Probability* = 100 %,   *Negative*
Probability = 0 %

Last, but not least, we see that government releases of small numbers of LeT personnel may in fact encourage LeT to attempt more attacks. In the case of the Indian government, it appears that these releases do not generate goodwill and in the case of the Pakistani government, these releases may allow LeT to return to business as usual.

TP-Rule AA-4. LeT carries out 1–2 attempted attacks one month after months in which:

- 0–2 LeT members were released from arrest.

Support = 24
Probability = 96 %,   *Inverse*   *Probability* = 100 %,   *Negative*
Probability = 0 %

On several occasions, Pakistan has arrested LeT members (and LeT leader Hafez Saeed in particular) in order to temporarily mollify an outraged international audience. In fact, the release of the LeT membership may be a subtle signal that the Pakistani military is encouraging further attacks by LeT. There were a few larger-scale releases, primarily by Pakistani authorities, after the Mumbai 2008 attack. But these releases occurred after major crackdowns in which dozens were arrested. Pakistan remained under intense international scrutiny for months after the 2008 attacks so it is unsurprising that, despite the large-scale releases of arrested LeT personnel, Pakistan continued to restrict LeT activities and attempted attacks were kept to a minimum.

Figure 8.2 below summarizes the relationship between attempted attacks by LeT and the various variables described above.

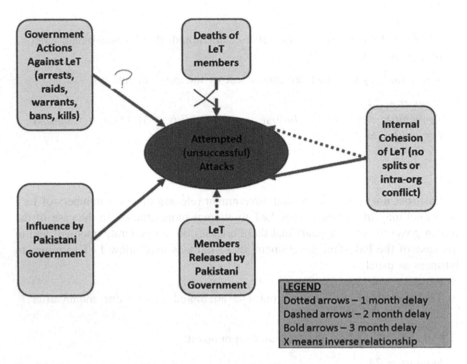

Fig. 8.2 Summary of variables related to attempted LeT attacks

8.3 Conclusion and Policy Options

In both types of attacks studied in this chapter—attacks on holidays and attempted attacks—it appears that when LeT is not blocked by government action it will at least attempt to carry out deadly attacks—mainly in India, but also Pakistan, Afghanistan, and Bangladesh.

This hypothesis is supported by the TP-rules showing that when no LeT leaders resigned or were arrested, LeT was able—within a couple of months—to attempt attacks including unsuccessful ones as well as successful ones on holidays. This suggests that, at least as far as attacks on holidays and attempted attacks are concerned, cracking down on the LeT leadership is key to disrupting LeT's planning and operational stages.

The influence of the Pakistani government (often under the influence of Pakistan's powerful armed forces and the Inter Services Intelligence agency) is also critical. Though this has been recognized in the literature for many years, it has been difficult to justify, at least on the basis of open source information. The rules presented in this chapter leave little question that while Pakistani government action can be effective in reducing LeT attacks, the policy challenge has been inducing the Pakistani government to do so. A recent paper (Dickerson et al. 2011) suggests that this policy can be brought about by a combination of American and Indian direct pressure on the Pakistani government by coercive diplomacy and on LeT directly by covert action. Chapter 11 discusses a wider range of policy options that can be brought to bear on LeT.

References

BBC News. (2005). *Deadly blasts hit Indian capital*. October 29, 2005. http://news.bbc.co.uk/2/hi/south_asia/4388292.stm

Daily News & Analysis. (2007). Lashkar militant with explosives held on R-Day eve. January 25, 2007, PTI—http://www.dnaindia.com/india/report_lashkar-militant-with-explosives-held-on-r-day-eve_1076323 and SATP's chronologies of LeT activity—http://www.satp.org/satporgtp/countries/india/states/jandk/terrorist_outfits/lashkar_e_toiba.htm

Dickerson, J., & Mannes, A. (2011). *Subrahmanian VS (2011) Dealing with Lashkar-e-Taiba: A Multi-Player Game-Theoretic Perspective*. Athens: Proceedings of IEEE International Symposium on Open-Source Intelligence and Web Mining.

Rana, M. A. (2006). A to Z of Jihadi Organizations in Pakistan. Translated by Saba Ansan. Lahore: Mashal Books http://www.desistore.com/jehadiorg.html

Rediff.com. (2005). 6 militants storm Ayodhya, killed. July 5, 2005, http://in.rediff.com/news/2005/jul/05ayo.htm

South Asia Terrorism Portal. (2011). Incidents involving LeT. *Institute for Conflict Management New Delhi*. http://www.satp.org/satporgtp/countries/india/states/jandk/terrorist_outfits/lashkar_e_toiba_lt.htm

Tankel, S. (2011). *Storming the World Stage: The Story of Lashkar-e-Taiba*. London: C. Hurst & Co.

Chapter 9
Armed Clashes

Abstract This chapter describes the conditions under which LeT engages in armed clashes, usually with Indian local or national security forces. The chapter discusses several TP-rules about armed clashes involving LeT that were derived automatically from the LeT data set used in this book.

During the period of this study (over 20 years), LeT engaged in hundreds of armed clashes with security forces—primarily Indian in Jammu and Kashmir. An attack is a violent event where *intent* to cause harm to an individual, a class of individuals, an organization, or a state can be established. In contrast, an *armed clash* is a violent event which occurs when two parties meet, engage in a violent confrontation, but there was no evidence that LeT had *intent* to engineer the confrontation—either through an a priori or an a posteriori declaration or statement or through explicit evidence (e.g., interrogation of suspects, telephonic/voice/VoIP intercepts). Thus, an LeT ambush or an intentional attack by LeT on a military camp would count as an "armed attack" rather than an armed clash. It is possible that events classified as armed clashes were in fact carefully orchestrated LeT attacks—but were classified by us as armed clashes because of a lack of access to data that might indicate otherwise.

Three kinds of armed clashes are discussed in this chapter—armed clashes with *local security forces* (taken broadly to include city and state police and police forces) and *national security forces* (taken broadly to mean Indian and Pakistani military forces, and other national security services including organizations such as India's Intelligence Bureau—the Indian equivalent of the US Federal Bureau of Investigation). Rules were also generated for *unspecified armed clashes* where open source information did not specify LeT's adversaries in a clash.

In addition to the type of adversary, clashes were also grouped in terms of casualties on one or both sides.

The study identified literally *hundreds* of rules associated with armed clashes. The key variables that appear correlated (usually in combination with one another) with armed clashes are:

V. S. Subrahmanian et al., *Computational Analysis of Terrorist Groups: Lashkar-e-Taiba*,   137
DOI: 10.1007/978-1-4614-4769-6_9, © Springer Science+Business Media New York 2013

- *Trials of LeT Personnel.* The system automatically discovered TP-rules that provide compelling statistical evidence that within 1–2 months of periods when LeT personnel are tried—either in Indian or Pakistani courts or in international tribunals—armed clashes are likely to follow. While India in particular, but Pakistan and even Western countries, have occasionally prosecuted LeT operatives, there is little evidence that this reduces LeT operations in what was long its main theater, Jammu and Kashmir.
- *Government Action Against LeT Leaders and Personnel.* This study shows that Pakistani government action against LeT leaders and personnel leads to an *increase* in armed clashes within a period of 1–3 months. Specifically, arrests and kills of LeT personnel are followed by *armed clashes* (usually with Indian security forces) within a 1–3 month time frame. *This suggests that as far as LeT is concerned, violence may beget violence. Government targeting of LeT's leaders may not be an effective strategy for reining in terrorist acts by the group*, though this book will propose alternatives.
- *Internal Cohesion within LeT's ranks and Lack of Intra-Organizational Conflict* is correlated with armed clashes within a 1–3 month time frame. This suggests that promoting increased division, distrust, and suspicion within LeT's ranks through carefully targeted covert action offers a possible approach to diminishing operational effectiveness of LeT's personnel.
- *Messages of Sympathy and Identity.* When LeT launches a publicity campaign that uses sympathy, usually for Muslim victims worldwide and most often in Kashmir, and identity politics that focuses on Islam versus Hindus/Christians/Jews, there is a strong probability that armed clashes with Indian security forces will occur within 1 month. Messages of sympathy and identity are used by LeT to enrage and rally its supporters and Pakistani opinion, recruiting them to LeT's cause and subsequently using attacks to gain further sympathy for the shaheed (fallen).
- *Practicing Charitable Acts.* Like many terror groups around the world such as Hamas and Hezbollah, LeT has an extensive charitable arm. LeT was the first organized group to provide relief services after the 2005 Kashmir earthquake and also provided aid to the victims of Pakistan's 2010 floods. LeT also runs extensive hospital and ambulance networks. In conjunction with the previous bullet that draws on sympathy and identity politics, LeT aims to be a competent and honest provider of essential citizen services in a state whose officials are often neither competent nor honest.[1]
- *Alliances with Pakistani Security Forces and Non-State Armed Groups.* The TP-rules automatically learned from 21 years of monthly data about LeT provide

[1] As stated above, experts on LeT and south Asia who have done fieldwork in Pakistan have stated that some of these variables such as Pakistani military support for LeT and LeT's charitable activities are effectively constant conditions. In this study, these conditions were coded when a major media source mentioned them as occurring. The fact that major media cites this activity could indicate a trend that this activity was occurring at a greater level then at other times.

compelling statistical support for the hypothesis that alliances—not only with Pakistani security forces (which have long been known) but also with other non-state armed groups such as Jaish-e-Mohammed and Indian Mujahideen—provide LeT with financial, personnel, materiel, and intelligence resources to carry out operations resulting in armed clashes during periods of mobilization. Even though LeT's use as a proxy has long been suspected and/or confirmed during the recent *U.S. versus Tahawwur Rana* court transcripts, the role of alliances with non-state armed groups may come as a surprise. In conjunction with the previous bullet, this suggests that *increased coercive diplomacy and reduction of military and development aid* to Pakistan will adversely affect the ability of Pakistan's military to provide material support to both LeT and other non-state armed groups with whom they have a relationship.

- *Number of LeT Training Camps.* Finally, there is a strong correlation between the number of training camps that LeT has and armed clashes with Indian security forces. More training camps typically implies a larger set of trained militants who can be deployed for operations. Additionally, greater numbers of training camps implies greater support from the Pakistani military and thus an increased operational tempo.

9.1 Armed Clashes, Trials, and Tribunals

The data mining software derived a set of TP-rules that assert the existence of a relationship between a combination of messages of sympathy and identity in LeT's public statements and the occurrence of trials involving LeT members in Australia and the occurrence of armed clashes within 1–2 months between LeT and Indian security forces.

TP-Rule AC-1. LeT is involved in armed clashes with local security forces (with LeT casualties) 1 month after months in which:

- LeT had a campaign focused on sympathy and identity politics and
- LeT personnel were being tried in Australia.

Support = 10
Probability = 100 %,   *Inverse   Probability* = 90.9 %,   *Negative Probability* = 0 %

The most celebrated LeT trial in Australia involved Faheem Lodhi, a Pakistani immigrant to Australia, who trained in LeT camps and then plotted a kinetic attack on Australia's electricity grid. Lodhi is believed to have had contact (and perhaps been trained by) with Sajid Mir, an LeT/ISI operative believed to be in charge of LeT's foreign recruits (Rotella 2011; *ABC* 2004). Lodhi was discovered when the

French intelligence agency Direction de la Surveillance du Territoire (DST) tracked Willie Brigitte, a French convert to Islam known to harbor radical views, and discovered that he had traveled to Australia. They notified Australian intelligence, who found Lodhi while investigating Brigitte and realized he was buying bombmaking chemicals (Jacobsen 2009). Lodhi was arrested in 2003 and convicted in 2006. (*ABC* 2004) describes part of the transcripts of French judge Jean-Louis Bruguiere's interrogation of Brigitte.

This is an intriguing rule for what it reveals about LeT's priorities. In addition, throughout 2009, several individuals were on trial in Australia for plotting terrorist attacks. While such trials may not have played a major role in LeT decision-making, it was occurring in the period just after the 2008 Mumbai assault, a period of enormous international scrutiny on LeT. Thus, it would have been unsurprising if LeT had reduced operations in Jammu and Kashmir. Yet the opposite occurred. Kashmir was undergoing large-scale protests and LeT was lending its support through statements of sympathy for the suffering of the Kashmiris (Reuters 2009). LeT military operations in Jammu and Kashmir serve to continue to destabilize the region. This rule would indicate that targeting LeT operations worldwide does little to hamper LeT operations in the organization's primary theater of operations.

The TP-Rule (AC-2) below links ongoing trials and arrests of LeT members with additional armed clashes with Indian security forces.

TP-Rule AC-2. LeT is involved in armed clashes with local security forces (with LeT casualties) 2 months after months in which:

- 5–24 LeT personnel were arrested and
- LeT personnel were on trial in India or Pakistan.

Support = 15
Probability = 88.2 %,   *Inverse*   *Probability* = 75 %,   *Negative*
Probability = 26.3 %

This rule indicates that trials in India and Pakistan along with moderate levels of arrests do little to deter LeT operations in Jammu and Kashmir. There have been specific periods with more than 24 personnel arrested—there were periods of large-scale Pakistani crackdowns on LeT, particularly after the 2008 Mumbai assault.

9.2 Armed Clashes and Pakistani Government Action Against LeT Personnel

Arrests and tribunals/trials are non-violent means used to rein in LeT members. However, kinetic actions by governments (both India and Pakistan) can also be followed by more armed clashes—this is not to suggest a causal link that is impossible to verify, but rather a temporal link. The next set of TP-rules deal with kills of LeT personnel by government forces.

Membership in terrorist organizations can often be fluid. A member of one group may have deep links to another group. The founders of IM and SIMI, many of whom trained with LeT are an example. These conduits between groups can extend a group's resources and capabilities. For example, in 2009 Mufti Obaidullah, a top leader of the Indian Islamist group the Asif Reza Comando Force, who also had close ties to LeT, was arrested. He told authorities that he had run a safehouse in Bangladesh that was used to infiltrate LeT operatives into and out of Kashmir (Dhaka Mirror 2009).

TP-Rule (AC-3) provides support for two hypothesis: first, that sharing membership with other non-state armed groups extends LeT's capabilities, allowing them to continue to carry out attacks by leveraging the capabilities of other groups. The second hypothesis is that offensive government action against LeT is less effective when LeT is allied with other non-state armed groups as these actors provide it with a capacity to absorb systemic shock. However, as most LeT members are killed in armed clashes, it is possible that the continuing armed clashes reflect periods of high infiltration.

TP-Rule AC-3. LeT is involved in armed clashes with local security forces (with LeT casualties) one month after months in which:

- LeT members were also members of other non-state armed groups and
- 0–8 LeT members were killed by the government (usually India).

Support = 11
Probability = 91.7 %,   *Inverse*   *Probability* = 100 %,   *Negative*
Probability = 0 %

Another TP-rule automatically derived by our system further supports this hypothesis.

TP-Rule AC-4. LeT is involved in armed clashes with local security forces (with LeT casualties) 1 month after months in which:

- LeT members were also members of other non-state armed groups and
- 5–61 LeT members were arrested by the Indian or Pakistani government.

Support = 10
Probability = 100 %,   *Inverse*   *Probability* = 90.9 %,   *Negative*
Probability = 33.3 %

This TP-rule further supports the hypothesis of TP-Rule (AC-3), suggesting that limited government action against LeT members does little to prevent or deter continuing LeT operations, particularly when LeT members also belong to other terrorist groups and thus have access to a broader base of resources.

9.3 Armed Clashes and Internal Cohesion of LeT

The internal cohesion of LeT is critical to their ability to function—LeT functions as a potent, lethal force when there is no factionalization amongst its leadership ranks. This same property applies also when it comes to armed clashes with Indian security forces. When LeT has not splintered, they appear to be more capable of planning complex operations—and as a consequence, operations in Jammu and Kashmir will continue unabated.

The TP-rule below shows what happens when LeT is not only internally cohesive, but also allied with Pakistan's security forces.

TP-Rule AC-5. LeT is involved in armed clashes with local security forces (with LeT casualties) 2 months after months in which:

- LeT is allied with Pakistan's security forces
- LeT is not splintering

Support = 11
Probability = 100 %,   *Inverse*   *Probability* = 100 %,   *Negative*
Probability = 0 %

9.4 Armed Clashes and Charitable Acts by LeT

Just as Hamas is embedded within the fabric of Palestinian society and Hezbollah is embedded within the fabric of Lebanese Shia society, LeT provides an array of social services ranging from earthquake relief to schools, to medical care. As stated above, LeT values jihad (holy war) and dawa (preaching) equally. For LeT, social services are a critical method for spreading the organization's methods and bringing more Pakistanis to embrace its version of Ahl Hadith Islam.

The *effect* of LeT's charitable operations is clear:

- LeT is now a full-fledged "essential service" provider within the Pakistani state. Due to incompetence and corruption, the Pakistani state does not consistently provide essential services to its citizens. LeT has stepped into the vacuum providing services with high competence and without asking for bribes. This provision of social services gives ample opportunity for LeT to present its message and win the allegiance of Pakistan's citizens. Pakistan's public school system is ineffective (Dawn.com 2011) and families seeking a low cost alternative often turn to LeT run schools, in which the tuition is subsidized for needy families. The education in LeT schools includes modern subjects but also a large dose of LeT's ideology. Doctors affiliated with LeT are instructed to stay in touch with patients and proselytize as they provide care.

• Ripping LeT out of the fabric of Pakistani society will be exceedingly difficult and will be particularly resistant to counter-radicalization efforts based outside of the country. But the Pakistani government would need to fill the vacuum created when LeT services are removed, something it has shown neither the desire nor the competence to do.

Rule (AC-6) below illustrates this point precisely. It states that when LeT is a charitable organization and when between 6 and 24 members of LeT are arrested, the group is involved in an armed clash within two months (with LeT casualties).

This rule—in addition to supporting the hypothesis that LeT's charitable acts may be connected to future armed activities by the group—also further supports the earlier hypothesis that arrests of LeT members have little efficacy in reducing LeT operations.

TP-Rule AC-6. LeT is involved in armed clashes with local security forces (with LeT casualties) 2 months after months in which:

• LeT was practicing as a charitable organization and
• 6–24 LeT members were arrested by the Indian or Pakistani government

Support = 11
Probability = 91.7 %,   *Inverse   Probability* = 84.6 %,   *Negative Probability* = 20 %

TP-Rule AC-7. 0–13 members of LeT are killed in Jammu and Kashmir three months after months in which:

• LeT provides social service medical programs and
• LeT is not banned in Pakistan.

Support = 16
Probability = 88.9 %,   *Inverse   Probability* = 100 %,   *Negative Probability* = 0 %

The above rule indicates that when LeT is permitted to go about its usual business, i.e., providing social services in Pakistan, and it is not banned by the government, then it is likely to be conducting operations in Jammu and Kashmir.

9.5 Armed Clashes and Deaths of LeT Leaders

Whether looking at 1, 2, or 3-month offsets, it appears that LeT engages in armed clashes after the death of some of its leadership. This raises important issues, as the effectiveness of the decapitation strategy (that is, killing the leaders of terrorist groups) is hotly debated among counterterror specialists (Jordan 2009, 2011). Rule (AC-8) below indicates when only a single battlefield commander was killed, LeT operations continue at least into the next month.

TP-Rule AC-8. LeT is involved in armed clashes with local security forces (with LeT casualties) 1 month after months in which:

• LeT members were also members of other non-state armed groups and
• 0–1 LET leaders were killed.

Support = 10
Probability = 100 %,   *Inverse   Probability* = 90.9 %,   *Negative*
Probability = 33.3 %

This TP-rule provides further powerful evidence that LeT is involved in armed clashes after LeT commanders die. Yet another similar rule applies to the case of a 2-month time offset.

TP-Rule AC-9. LeT is involved in armed clashes (with LeT casualties) two months after months in which:

• LeT leaders were killed.

Support = 58
Probability = 86.6 %,   *Inverse   Probability* = 78.4 %,   *Negative*
Probability = 40 %

This same TP-rule also applies—with exactly the same statistics—when the number of LeT commanders who died in a given month ranged between 1 and 7, indicating that the number of LeT commanders killed has no real impact, at least in the short term, on LeT operations.

9.6 Armed Clashes and Support by the Pakistani Government

It will come as no surprise to most readers that the data mining software automatically discovered TP-rules linking support by the Pakistani military and deaths of LeT personnel in Jammu and Kashmir. This connection has been widely reported in the press and has been the subject of much speculation (Clarke 2010).

TP-Rule AC-10. 0–13 members of LeT are killed in Jammu and Kashmir three months after months in which:

- LeT receives military support from the Pakistani government.

Support = 15
Probability = 93.8 %,   *Inverse*   *Probability* = 100 %,   *Negative*
Probability = 0 %

Usually, LeT deaths in Jammu and Kashmir are in encounters with security forces, so this rule is probably closely related to armed clashes and supports the hypothesis that military support from the Pakistani government is critical to enabling LeT operations in Jammu and Kashmir.

9.7 Armed Clashes and Number of LeT Training Camps

LeT has an extensive network of training camps, primarily located in Pakistani-controlled Kashmir. This shows that LeT is at least tolerated and allowed to function by the Pakistani government, and at worst, that it is explicitly supported by them. These camps, primarily located near Muzzafarabad (the capital of Pakistani controlled Kashmir), are where LeT operatives undergo their extensive training before infiltrating into the Indian province of Jammu and Kashmir. There are a few major training camps, but LeT opens smaller camps or launching bases close to the border prior to major infiltration efforts. The system automatically derived TP-rules that showed a relationship between armed clashes involving LeT and the presence and number of such training camps.

TP-Rule AC-11. LeT is involved in armed clashes (with LeT casualties) one month after months in which:

- LeT had between 1 and 23 training camps.

Support = 31
Probability = 86.5 %,   *Inverse   Probability* = 97 %,   *Negative*
Probability = 33.3 %

This rule has substantial support and shows a strong relationship between LeT involvement in armed clashes and presence of a large number of training camps. The relationship itself is unsurprising—training camps train mujahideen to go out and reconnoiter targets, to carry out surveillance operations and practice runs, and it is inevitable that increased training activity is correlated with future operations. Another similar rule reinforces this point.

TP-Rule AC-12. LeT is involved in armed clashes (with LeT casualties) one month after months in which:

- LeT had between 1 and 25 training camps and
- 0–11 LeT members were killed by the government.

Support = 32
Probability = 91.2 %,   *Inverse   Probability* = 93.9 %,   *Negative*
Probability = 33.3 %

These two TP-rules, together with others of the same ilk, provide strong evidence that LeT's operations increased when they were permitted to run training camps, which allow them to expand their armed cadres and infiltrate them into Jammu and Kashmir where they engage in armed clashes with Indian security forces.

9.8 Conclusion and Policy Options

This chapter shows that LeT is engaged in armed clashes a few months after situations when certain variables are true. Figure 9.1 shows these dependencies succinctly.

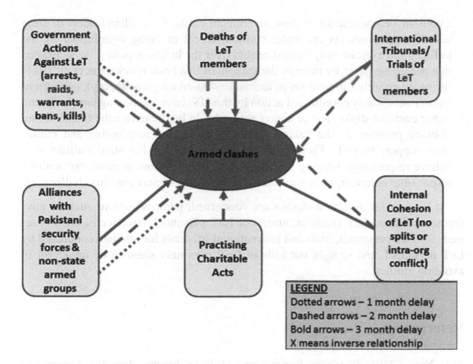

Fig. 9.1 Summary of variables related to armed clashes involving LeT

As in the cases of previous chapters, Fig. 9.1 identifies some policy options to reduce armed clashes involving LeT. However, unlike explicit actions intentionally undertaken by LeT discussed in preceding chapters, armed clashes are a result of unplanned encounters between LeT personnel and (usually) Indian security forces. They suggest pre-meditation by LeT (otherwise their armed personnel would not be operating in Jammu and Kashmir), but the purpose of these "movements" is not clear.

Figure 9.1 immediately suggests several important policies.

- *Disrupting LeT's Internal Cohesion.* It appears in Fig. 9.1 that LeT's internal cohesion is key to its ability to plan operations, which results in the movement of armed personnel who, in turn, get involved in armed clashes. Thus, disrupting LeT's ability to function as a single, cohesive organization could be key to disrupting LeT armed operations.
- *Disrupting LeT's Finances.* With the looming end of the US-led war against al-Qaeda in Afghanistan, the US does not need Pakistan with the same urgency as it did in the first decade of the 21st century. Reducing US military and development aid to Pakistan will reduce Pakistan's ability to provide financial, military, and materiel support to LeT.
- *Increasing the cost of doing business for the Pakistani military.* The Pakistani military establishment rules Pakistan with an iron fist. Even though the façade of

a civilian, democratically elected government exists, the civilian leaders of these so-called governments are under constant threat of being overthrown by the military and exercise only limited control over the levers of policy. As suggested in a preceding paper by three of the authors of this book (Dickerson et al. 2011) (using an entirely different set of techniques based on game theory), concurrent but not necessarily coordinated action by the US (covert action against LeT) and either coercive diplomacy or covert action by India, increases the financial and political pressure on the Pakistani military to behave responsibly and reduce their support to LeT. There is no guarantee that the Pakistani military will behave responsibly, keeping the interests of their citizens in mind, but without responsible behavior, efforts to rein in LeT's harmful acts remains challenging.

At least as far as armed clashes are concerned, punitive actions such as government raids on LeT facilities, arrests of LeT personnel, issuance of arrest warrants for LeT personnel, trials and international tribunals for offenses conducted by LeT personnel, and straight out kills of LeT personnel seem to be unhelpful in avoiding clashes.

References

ABC News (2004). *Transcript: Interrogations of Willie Brigitte*, Australian Broadcasting Corporation Four Corners program, http://www.abc.net.au/4corners/content/2004/s1131449.htm

Clarke R (2010). "Lashkar-i-Taiba: The Fallacy of Subservient Proxies and the Future of Islamist Terrorism in India," *The Letort Papers*, Strategic Studies Institute, U.S. Army War College, March 2010

Dawn.com (2011). Education Emergency Pakistan. Retrieved March 9, 2011, from http://www.dawn.com/2011/03/09/education-emeregency-pakistan.html

Dhaka Mirror (2009). Foreign militants used country as transit point. Retrieved September 2, 2009, from http://www.dhakamirror.com/headlines/foreign-militants-used-country-as-transit-point/

Dickerson, J., Mannes, A., & Subrahmanian, V. S. (2011). Dealing with Lashkar-e-Taiba: A multi-player game-theoretic perspective. *Proceedings of the IEEE International Symposium on Open-Source Intelligence and Web Mining*. Athens, Greece, September 2011

Jacobsen, G. (2009). Five guilty in Sydney terrorism trial. *Sydney Morning Herald*. Retrieved October 16, 2009, from http://www.smh.com.au/national/five-guilty-in-sydney-terrorism-trial-20091016-h06l.html

Jordan, J. (2009). When Heads Roll: Assessing the Effectiveness of Leadership Decapitation. *Security Studies*. http://cpost.uchicago.edu/pdf/Jordan.pdf

Johnston, P. (2011). Assessing the Effectiveness of Leadership Decapitation in Counterinsurgency Campaign. *International Security*. Retrieved July 2011, from http://patrickjohnston.info/materials/decapitation.pdf

Reuters (2009). No let-up in IHK protests. Reuters, Retrieved June 2, 2009, from http://www.nation.com.pk/pakistan-news-newspaper-daily-english-online/Politics/02-Jun-2009/No-letup-in-IHK-protests

Rotella, S. (2011). Pakistan and the Mumbai Attacks: The Untold Story, *Amazon Kindle book*

Chapter 10
Computing Policy Options

Abstract This chapter describes the methodology and the algorithm used to automatically compute policy options. It provides a mathematical definition of a policy against LeT and then proves the LeT Violence Non-Eliminability Theorem that shows there is no policy that will stop all of LeT's terrorist actions. The reason for this is that attacks on holidays are carried out in situations that are inconsistent with situations when LeT carries out other types of attacks. The chapter presents an algorithm to compute all policies (in accordance with the mathematical definition of policy) that have good potential to significantly reduce all types of attacks carried out by LeT (except for attacks on holidays). Readers who do not wish to wade through the technical details can skip directly to Sect. 10.5 which summarizes the results of this chapter.

Chapters 4–9 of this book describe the conditions under which LeT carries out six types of terrorist attacks (or attempted attacks). The variables that are related to these attacks are partially summarized in the figures shown at the end of each of these chapters. Although these summary figures do suggest possible policies to mitigate one type of attack, they do so at the risk of aggravating another kind of attack. Deriving "good" policies requires a process of simultaneously considering how the wide variety of attacks carried out by LeT might be mitigated.

In this chapter, we briefly describe the computational methods used to generate policies automatically from the TP-rules we have described and discussed in Chaps. 4–9. Appendix B summarizes the set of all TP-rules described in this book. Our policy computation algorithms build upon the methods to use linear programming to compute all minimal models of logic programs developed by one of the authors in a series of papers (Bell et al. 1994a, b).

V. S. Subrahmanian et al., *Computational Analysis of Terrorist Groups: Lashkar-e-Taiba*,   149
DOI: 10.1007/978-1-4614-4769-6_10, © Springer Science+Business Media New York 2013

10.1 Policy Analysis Methodology

In this section, we briefly describe the methodology we used to automatically generate policies to significantly rein in LeT's violent activities. Our methodology used several steps that are outlined below.

- **Step 1—Rule Elimination:** We only considered TP-rules presented in this book that describe conditions when LeT carries out one of the kinds of attacks described here. In particular, we eliminated the rules that describe conditions when LeT does *not* carry out such attacks. This left a total of 53 rules.
- **Step 2—Time Offset Elimination:** As all rules in this book have time offsets of 1, 2, or 3 months, we eliminated all the time offsets from the TP-rules. The resulting set of "reduced rules" all have the form

$$A(i) \leftarrow B_1 \& \ldots \& B_n$$

  where $A(i)$ describes an action taken by LeT with intensity level i and $B_1 \& \ldots$ & B_n is a conjunction of environmental literals (atoms or negated atoms). Intuitively, this rule can be read now as: LeT will take action A with intensity level i (with high confidence) within 1–3 months of months when $B_1 \& \ldots \&$ B_n is true.
- **Step 3—Environmental Variable Constraints:** We examined the bodies ("if part") of the 53 TP-rules presented in this book (after Step 1) and found that some of these rule bodies involve environmental atoms that cannot be easily changed. For instance, the environmental variable describing LeT as a religious organization is one we deemed unchangeable. It is not realistic to believe that LeT can be persuaded to give up its Ahl Hadith religious persuasion. Along the same lines, we deemed the variable describing LeT's territorial claims on Kashmir unchangeable—this is not a variable US or Indian policy can expect to reasonably affect.
- **Step 4—Policy Computation:** Using the remaining rules, we developed methods to compute the set of all policies (i.e., ways to change the values of environmental variables, such as to the constraints introduced in Step 2) that prevent LeT from taking any harmful action.

Thus, we now have 53 rules, each represented in the above form. We call this the LeT rule base (**LeTRB**)—the set of rules from which we want to derive policies. The technically savvy reader will note that these rules are "propositional logic" rules even though they appear to be first order rules because no variables appear in any of the TP-rules presented in this book (Mendelson 2009).

10.2 Computing Policies

Suppose *Body*(**LeTRB**) is the set of all literals (positive and negative) that occur in the body of any rule in **LeTRB**.

Recall, as usual, that a pair of literals (*L*, ~ *L*) are called *complementary literals*. For instance, *kill_LeT_leaders* and ~ *kill_LeT_leaders* is a complementary pair, and each of these two literals is the complement of the other one.

Let *CompBody*(**LeTRB**) be the set of all literals { ~ *L* | *L* is in *Body*(**LeTRB**)}. In other words, *CompBody*(**LeTRB**) is the complement of every literal that occurs in the body of any of the rules in **LeTRB**.

Formally, a *policy P that potentially eliminates all violent acts*[1] *of LeT* (or just *policy*, for short, as it is understood that the goal of this book is to rein in LeT) is a consistent subset of *CompBody*(**LeTRB**) that satisfies three conditions:

1. For each rule *r* in **LeTRB**, there is a literal in *P* whose complement also occurs in the body of rule *r*.
2. *P* must satisfy all specified environmental variable constraints (Step 3 above).[2]
3. There is no *strict subset P'* of *P* satisfying the preceding two conditions.

Intuitively, a policy is a set of literals denoting actions that any organization wanting to "rein in" LeT might want to consider taking. From a technical perspective, a policy can be viewed as a "minimal model" of a logic program (Minker 1982) that cannot make the head of any rule true and that must, in addition, satisfy environmental variable constraints. Minimal model computations have been studied extensively over the years, and a variety of fast algorithms to find them have been produced (Bell 1994a, b; Subrahmanian et al. 1995).

Structurally, a policy must satisfy several requirements:

1. First, a policy must only consist of literals in *CompBody*(**LeTRB**). Note that each literal in *CompBody*(**LeTRB**) prevents at least one rule in **LeTRB** from firing (and possibly more than one).
2. Second, a policy must be consistent—it cannot contain both *L* and ~*L* for any literal *L*.
3. Third, a policy must prevent each rule in the **LeTRB** rule base from firing. Thus, a policy ensures that there is *no way* for LeT to carry out any of the violent acts that they have carried out in the past.
4. Fourth, a policy must be *minimal*. It must not recommend more actions to be taken by appropriate decision makers than are strictly necessary.

Our first result is the following theorem, which we can prove mathematically.

---

[1] Only violent acts considered in this book are included here.

[2] In other words, if we stated that a specific literal cannot possibly be made true (e.g., the variable which says LeT stops being a religious organization cannot possible be made true), then this must be respected.

The LeT Violence Non-Eliminability Theorem. *There is no policy that potentially eliminates the violent acts by LeT* discussed in Chaps. 4–9.

Proof sketch. Suppose $\mathbf{LeTRB} = \{r_1,...,r_n\}$ and let $cl_i = \{ \ L \mid L$ is a literal in the body of rule $r_i\}$. Thus, each cl_i is a *logical clause* in the strict sense of (Robinson 1965). The resulting set of clauses, $\mathbf{CL(LeTRB)} = \{ \ cl_i \mid 1 \leq i \leq n\}$ is logically unsatisfiable (Robinson 1965).

An alternative proof proceeds via the concept of a *hitting set*. If we consider the sets $\{cl_i \mid 1 \leq i \leq n\}$, there is no hitting set of this set that is consistent.

This theorem explains, for the very first time (in a technical way), why dealing with LeT is such a hard problem. Whatever policies we adopt may have some negative effects.

The good news, however, is that if we ignore Rules (AOH-1) and (AOH-4) which deal with attacks on holidays, we can formally compute policies that could potentially eliminate all violent acts by LeT considered in this book. We use **LetRedRB** to denote the reduced version of **LetRB** that eliminates these two rules, and computer policies using **LeTRedRB**. Chapter 11 describes what these policies are and discusses them in detail, but the following section describes how we computed these policies.

10.3 Computing Policies to Potentially Eliminate (Most) Violent Acts by LeT

We use integer linear programming to solve the problem of violating all violent acts carried out by LeT (that are studied in this book) with the exception of some attacks on holidays.

Let *Literals* be the set of all literals L that appear anywhere in any rule (head or body) in **LetRedRB**. Let X_L be a variable (whose value is unknown) telling us if literal L is included in a policy or not. We now define a set of linear constraints $LC(LeT)$ as follows.

1. For each rule $A \leftarrow B_1 \ \& \ .... \ \& \ B_n$ in **LeTRedRB**, $LC(LeT)$ contains the constraint $X_A + \sum_{i=1}^{n} (1 - X_{B_i}) \geq 1$. This constraint says that either A is true or one of the B_i's is false.
2. For each environmental variable constraint saying that the value of a particular environmental atom A is set to a fixed value c *(0 denoting false, or 1 denoting true)*, $LC(LeT)$ contains the constraint $X_A = c$.
3. For each pair of complementary literal L, $\sim L$, $LC(LeT)$ contains the constraint $X_L + X_{\sim L} \leq 1$ This constraint says that at most one of L or $\sim L$ is true.
4. If A occurs in the head of any rule in **LeTRedRB**, then $LC(LeT)$ contains the constraint $X_A = 0$. This constraint says that we would like to ensure that LeT cannot carry out any violent activities.

5. *LC(LeT)* contains a constraint $X_L \in \{0, 1\}$ indicating that all variables are either 0 or 1.

The following important theorem tells us that we can find policies that potentially prevent LeT from carrying out any of the violent acts for which we derived TP-rules in Chaps. 4–9 (excluding AOH-1 and AOH-4).

Theorem. *LC(LeT)* is solvable.

The following algorithm tells us how to generate *all* policies that potentially prevent LeT from carrying out any of the violent acts for which we derived TP-rules in Chaps. 4–9 (except for AOH-1 and AOH-4).

The Policy Computation Algorithm works as follows.

- It solves the integer linear program shown at the beginning of the while loop using any standard integer linear program solver.
- If the integer linear program is solvable, then it has found a set *S* of literals in the rule bodies which, if negated, would prevent every rule in **LetRedRB** from firing. The set $\{ \sim L \mid L \text{ is in } S \}$ is therefore a policy. It adds this to the set of policies found thus far. Moreover, to ensure that no more policies that are supersets of this one are found by the algorithm, it adds a constraint to the set of constraints.
- If the integer linear program is unsolvable, then we have found all policies that potentially eliminate violent attacks by LeT.

POLICY COMPUTATION ALGORITHM

Policies = { }; (* no policies found so far *)
Solvable = true;
Constraints = *LC(LeT)*;

While Solvable **do**
$\quad$ S = **minimize** $\sum_{L \text{ occurs in the body of a rule in letRedRB}} X_L$;
$\quad\quad$ **Subject to** *Constraints*;

$\quad$ **If** a solution exists **then**
$\quad\quad$ Policies = Policies $U \{ \sim L \mid L \text{ is in } S \}$;
$\quad\quad$ Constraints = Constraints $U \{ \sum_{L \in S} X_L \leq card(S) \}$
$\quad$ **Else** Solvable = false;

Return Policies

Theorem. The Policy Computation Algorithm finds all policies that potentially eliminate violent attacks by LeT.

We ran the Policy Computation Algorithm and generated a total of 8 unique policies in all that potentially eliminate violent attacks by LeT. Four of these policies vary from the other four only in very minor technical details—hence, in the Chap. 11, we will discuss four policies in great detail.

10.4 A Note on Alternate Policy Computation Methodologies

In this section, we discuss the intuition behind the formulation of the Policy Computation Algorithm given above and compare it to other possible (equivalent) approaches. We also briefly overview some directions one could take to further improve the theoretical and empirical power of this algorithm.

We use an integer linear programming approach. This decision was motivated by both the simplicity and generality of the formulation and, importantly, the availability of efficient software packages to solve such problems. The industry standard integer programming software suites—IBM's ILOG CPLEX or the Gurobi Optimizer—are capable of solving incredibly complicated integer programs through a mix of highly specialized pre-solving, customized cut generation, and efficient tree search.

Still, those familiar with optimization (and complexity theory in computer science) will recognize that our problem is susceptible to other formulations. For example, the basic hitting set calculation is reducible to simply solving a variant of the boolean satisfiability problem, more colloquially known as SAT. (Cook 1971; Karp 1972) provide excellent descriptions of complexity reductions. Efficient, but less industrially honed, SAT solvers exist; currently, the academic leaders include the serial MiniSAT (Eén and Sörensson 2004) or the parallel ManySAT (Hamadi et al. 2009), both of which can solve SAT formulae with millions of variables. These could have been used to produce identical results to those in this chapter and the next.

Extending the model. One advantage to our integer programming approach is the ability to easily add *weights* to the variables X_L representing whether or not a literal L is included in a policy. Intuitively, if a literal L (say, *kill_LeT_leaders*) is given a large weight w_L, then that literal is "hard" for a policy maker to implement in reality. Similarly, if a literal L' (say, *no_government_ban*) has a smaller weight $w_{L'}$, then it is "easier" to effect change relative to that literal in reality. Given a mapping of literals to weights, we could then—through the addition of these weights to our objective function—minimize the overall *weight* of active variables rather than the overall cardinality. This would allow policymakers to use their real-world knowledge of the intricacies of diplomacy to personalize these automatically generated policy recommendations.

Scaling to larger rule bases. In this book, we intensely explored a set of 61 rules pertaining to violent attacks performed by LeT. One could imagine, however, modeling a much broader situation (e.g., all terrorist groups in the region or the world), resulting in an enormous set of automatically generated rules. For these larger problem instances, the computational complexity of computing optimal policies (either by solving the integer linear program or otherwise) becomes overwhelmingly large. In these experimentally intractable cases, one could find approximate solutions—although, as we now discuss, this is a theoretically difficult problem.

The hitting set problem is, intuitively, closely related to the well-studied NP-Complete set covering problem (in fact, the problem is *equivalent* to set cover through a simple reduction using bipartite graphs). It has been shown that the set covering problem is inapproximable in theory (Lund and Yannakakis 1994). In fact, the best any approximation algorithm can do in theory is roughly equivalent to the standard greedy algorithm for approximating set cover and hitting sets, which can be off from optimal by a log factor of the number of rules considered (Alon et al. 2006). However, in practice, such approximation algorithms often lead to very good—and, importantly, tractable—approximate results. Furthermore, being able to approximately solve very large models could outweigh being limited to smaller models, even if these small models could be solved to optimality.

Dealing with infeasibility. Actors on the world stage are self-interested but rarely rational, at least in the game-theoretic sense. As we saw in this chapter (specifically with rules AOH-1 and AOH-4), this lack of rationality may lead to unpredictable or seemingly contradictory actions, which, in turn, leads to infeasibility of the hitting set problem. In the event of infeasibility, a policy analyst could relax the objective function to instead automatically calculate the *largest* set of rules that could possibly be prevented from firing, rather than requiring that *all* rules do not fire. These are closely related to the concept of *maximally consistent subsets* of logic programs (Baral et al. 1992). A technique to do this using integer linear programming is suggested in (Bell et al. 1994b). This problem, at least without variable weights, is also equivalent to the maximum boolean satisfiability problem (or MAX-SAT), another well-known NP-Hard problem. Expert knowledge could also be used to cull certain rules from the larger rule base.

10.5 Conclusion

Policy makers, security analysts, and military decision makers in numerous countries have long grappled with the problem of eliminating violent acts by Lashkar-e-Taiba. Despite the collective wisdom over a period of 20 years, little has been accomplished as far as reining in LeT is concerned. One may argue, in fact, that LeT has grown stronger during this time, carrying out increasingly bold and deadly violent acts such as the Mumbai attacks.

In this brief chapter, we describe our policy analytics methods that merge important elements of mathematical logic, logic programming, and integer linear programming. We apply this policy analytics methodology to computing strategies that may help rein in LeT.

Our first major result is the LeT Violence Non-Eliminability Theorem which shows mathematically, for the first time, that external actors (e.g., the US or India) cannot eliminate all violent acts by LeT (unless LeT changes its behavior and acts in accordance with different rules).

The major reason for this is that the conditions under which LeT carries out attacks on holidays (and ways to reduce LeT attacks on holidays) are inconsistent with the conditions under which LeT carries out other types of attacks. We show that if we ignore two rules derived in Chaps. 4–9 dealing with LeT attacks on holidays, we can find policies that potentially eliminate all other types of LeT-backed attacks studied in Chaps. 4–9.

We then develop and describe the Policy Computation Algorithm to find all such policies, leveraging an algorithm due to (Bell et al. 1994a, b). That algorithm, used to compute "minimal models" of logic programs, is adapted to compute policies.

We conclude with some caveats: the policies generated by our algorithm only *potentially* eliminate LeT attacks. They do not guarantee them. What they do guarantee is that the policies generate an environment "around" LeT that is maximally conducive to reduced attacks.

We describe the policies generated by our Policy Computation Algorithm in detail in the Chap. 11.

References

Alon, N., Moshkovitz, D., & Safra, S. (2006). Algorithmic construction of sets for k-restrictions. *The ACM Transactions on Algorithms, 2*(2), 153–177.

Baral, C., Kraus, S., Minker, J., & Subrahmanian, V. S. (1992). Combining knowledge bases consisting of first order theories. *Computational Intelligence, 8*(1), 45–71.

Bell, C., Nerode, A., Ng, R., & Subrahmanian, V. S. (1994a). Implementing deductive databases by mixed integer programming. *ACM Transactions on Database Systems, 21*(2), 238–269.

Bell, C., Nerode, A., Ng, R., & Subrahmanian, V. S. (1994b). Mixed integer methods for computing non-monotonic deductive databases. *Journal of the ACM, 41*(6), 1178–1215.

Cook, S. (1971). *The complexity of theorem proving procedures, Proceedings of the third annual ACM symposium on Theory of computing.* pp. 151–158.

Eén, N., & Sörensson, N. (2004). An extensible SAT-solver. In *Theory and applications of satisfiability testing* (pp. 333–336). Berlin: Springer.

Hamadi, Y., Jabbour, S., & Sais, L. (2009). ManySAT: A parallel SAT solver. Int. Journal on Satisfiability, Boolean Modeling and Computation (JSAT), 6, 245–262 IOS Press.

Karp, R. (1972). Reducibility among combinatorial problems. In R. E. Miller & J. W. Thatcher (Eds.), *Complexity of computer computations* (pp. 85–103). New York: Plenum.

Mendelson, E. (2009). *Introduction to mathematical logic* (5th ed.). London: Chapman and Hall/CRC Press

Minker, J. (1982). *On indefinite databases and the closed-world assumption, Proceedings of 6th international conference on automated deduction,* Lecture Notes in Computer Science Vol. 138, pages 292–308

Lund, C., & Yannakakis, M. (1994). On the hardness of approximating minimization problems. *Journal of the ACM, 41*(5), 960–981.

Robinson, J. A. (1965). A machine-oriented logic based on the resolution principle. *Journal of the ACM, 12*(1), 23–41.

Subrahmanian, V. S., Nau, D. S., & Vago, C. (1995). WFS + branch and bound = stable models. *IEEE Transactions on Knowledge and Data Engineering, 7*(3), 362–377.

Chapter 11
Policy Options Against LeT

Abstract The Policy Computation Algorithm described in Chap. 10 was used to generate a total of 8 policies that have a reasonable chance of reducing most types of LeT backed attacks. This chapter describes 4 of these policies in detail (the other 4 are very similar). It shows that these policies are complex, involving many different actions to be taken—yet they overlap extensively in terms of what should be done (and what should not be done) in combating LeT terror acts. The chapter also includes some tactics that policy makers may consider in implementing these policies.

Chapter 10 describes methods to automatically compute policies against LeT which have the potential to prevent the types of attacks studied in with the exception of attacks on holidays. These results do not mean that attacks will stop if the policies described in this chapter are implemented—the results are based on the TP-rules discovering behavioral models of LeT based on their *past* behavior. Successful terrorist groups are extremely adept at changing their behavior in response to counter-terror strategies (Ganor 2005) and policies must be constantly reshaped to account for changes made by adversaries.

Figure 11.1 below shows a summary of the TP-rules discussed in Chaps. 4–9. The bold blue arrows indicate an apparent positive relationship between months when the environment satisfies the condition mentioned in the blue boxes and subsequent months (1–3 months later) when LeT takes the action shown in the red boxes. The dotted brown arrows indicate a negative relationship.

This chapter describes the set of policies computed automatically from the TP-rules derived in Chaps. 4–9 (subject to the policy generation methodology described in Chap. 10) using the Policy Computation Algorithm (PCA) described in the preceding chapter. The PCA also identified a total of eight policies, but for all practical purposes, these policies boil down to four policies—the other four are minor variants of the ones presented here.

V. S. Subrahmanian et al., *Computational Analysis of Terrorist Groups: Lashkar-e-Taiba*,   157
DOI: 10.1007/978-1-4614-4769-6_11, © Springer Science+Business Media New York 2013

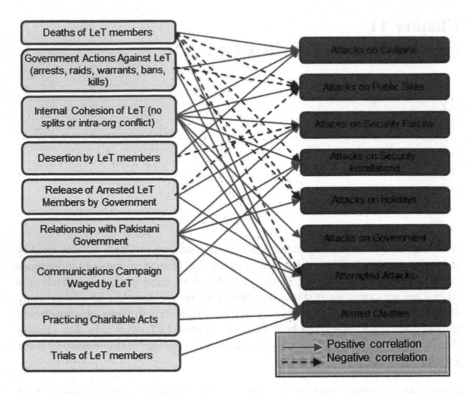

Fig. 11.1 Summary of relationships between LeT attacks and associated environmental variables

Table 11.1 Recommendations that are common to all policies generated by the policy computation algorithm to combat LeT

Actions to be performed in all policies	Actions not effective in all policies
Promote splintering of LeT	No government raids on LeT personnel
Promote intra-organizational conflict within LeT	No government arrests of LeT personnel
Disrupt Pakistani military support for LeT	No government arrest warrants for LeT personnel
Disrupt LeT's communication campaigns	No LeT personnel killed by government
Disrupt releases of arrested LeT personnel by Pakistani government	No mechanisms to encourage defection by LeT members
Encourage resignation of LeT leaders	
Disrupt LeT training camps	

Before discussing the details of these policies, the factors that were *common* to all of these policies are summarized. Table 11.1 provides a succinct summary of the *common parts* of these policies. The parts that are unique to each policy generated by the system are listed in Table 11.2.

Table 11.2 Actions unique to each policy against LeT generated by the policy computation algorithm—the complete policies represent the union of the acts shown here and those shown in Table 11.1

Policy ID	Additional actions to be taken or not taken
P1	Disrupt support provided by LeT to other Islamist organizations Ensure Pakistani government ban on LeT
P2	Ensure Pakistani government ban on LeT Reduce publicity generated by high profile trials of LeT personnel
P3	Ensure LeT does not provide support to other Islamist organizations Disrupt LeT's social services and medical programs
P4	Disrupt LeT's social services and medical programs Reduce publicity generated by high profile trials of LeT personnel

- *Disrupt the Internal Cohesion of LeT.* All the policies discovered by the Policy Computation Algorithm recommended taking steps that encourage splintering and intra-organizational conflict within LeT. These conflicts are sometimes linked to clashes between top LeT leaders, at other times they are *pro forma* splits to escape international sanctions or orchestrated by Pakistani intelligence. Whatever the cause of these lapses in internal cohesion, when LeT is engaged in infighting, either their interest in committing terrorist acts or their ability to conduct terrorist acts appears to be severely compromised. All three of these variables (promoting splintering, promoting intra-organizational conflict, and promoting the resignations of LeT leaders) are recommended by all of the policies found by the algorithm. The strategy of disrupting internal cohesion of LeT does not appear to have been suggested elsewhere in the literature; however some elements of this strategy were previously used to disrupt the functioning of other terror groups. The destruction of the Abu Nidal Organization is probably the best-known case of intelligence agencies eliminating a terrorist group by sowing discord within its ranks (Perry 1992), but many other terrorist groups have had difficulties maintaining internal cohesion (Shapiro 2007).
- *Encourage Resignations of LeT leaders.* The resignations of top LeT leaders go hand-in-hand with intra-organizational conflicts and organizational splits. In at least some of the cases, the resignation was because one element of LeT had been officially banned, and Hafez Muhammad Saeed, the leader was officially stepping down. These events often coincide with periods of international scrutiny and Pakistani governmental pressure on LeT to reduce its activities. Thus, stepped-up pressure by the international community has the desirable effect of causing an increased number of resignations of LeT leaders, which in turn are often followed by a reduction in the number of most terrorist acts committed by LeT. Whatever the reason, increased international pressure on Pakistan to reduce support for terrorism seems to be a good thing.
- *Disrupt LeT's communication campaigns.* As discussed above, LeT has an active publicity operation, which is used to drum up support for the organization, raise funds, and generate recruits. Every single policy generated by the

Policy Computation Algorithm required that LeT's public communications activities be disrupted, thus compromising LeT's ability to generate support within the local population. Other works in the literature discussing policy responses to LeT do not propose a disruption of LeT's communications campaigns as a possible policy toward LeT.

- *Prevent Releases of Arrested LeT personnel by Pakistan.* The TP-rules make it clear that releasing LeT prisoners largely serves as a powerful "shot in the arm" to LeT, energizing and reinvigorating them. This is particularly true when the personnel released are LeT leaders such as Hafez Saeed. Such releases may also signal an implicit "green light" from the Pakistani military that international pressure is off and that LeT is free to go ahead with violent operations.
- *Disrupt LeT training camps.* All the policies generated by the Policy Computation Algorithm required that steps be taken to disrupt LeT's training camps. While many of these camps are located throughout Pakistan (see Fig. 1.3), with a concentration in and around Lahore in the Punjab, there is also a large concentration of these camps in Pakistani controlled Kashmir. These camps provide the training and indoctrination to LeT recruits who subsequently carry out various kinds of terrorist acts.
- *Disrupt Pakistani military support for LeT.* All the policies discovered by the Policy Computation Algorithm recommend taking steps to disrupt Pakistani military support for LeT. This policy recommendation will come as little surprise to most analysts, who—with near unanimity—agree that LeT often acts as a proxy force for the Pakistani military.

In addition to the seven actions to be explicitly performed listed in Table 11.1, the Policy Computation Algorithm also recommended several actions that do not appear to reduce the likelihood of LeT carrying out violent operations. All of these actions (e.g., arresting and killing members of the terrorist organization) are standard operations for a government facing a terrorist organization, so it is strongly counter-intuitive that in the case of LeT, they not only have little effect, but seem to be followed by an uptick in subsequent terrorist action by LeT. Several possibilities present themselves. Many of the arrests, deaths, and defections of LeT personnel reflect actions by Indian security forces during periods of high-level violence in Jammu and Kashmir. During such periods, more LeT personnel would probably be killed because LeT was carrying out more attacks and infiltrations. Operationally, these are tactics and strategies that will continue to be used by Indian security forces. However, these findings highlight that traditional security responses to LeT are inadequate as long as LeT has an extensive infrastructure operating under state protection.

Table 11.1 shows the "common parts" of all the policies generated by the Policy Computation Algorithm. Table 11.2 below shows the actions that are unique to each policy generated.

The next four sections discuss each of these four policies in further detail. These sections also discuss the pros and cons of each of these policies.

11.1 Policy P1

Policy P1 requires taking all of the actions shown in Table 11.1—in addition, it requires performing two other actions.

The first action unique to policy P1 is that of disrupting the support provided by LeT to other Islamist organizations. There is evidence of a growing relationship between LeT and other organizations. For instance, LeT is believed to have or had ties to:

- *Al-Qaeda.* For instance, Abu Zubaydah, al-Qaeda's chief of operations after the death of Mohammed Atef in an air strike in Afghanistan, was caught in an LeT safe house in Faisalabad (Wall Street Journal 2008). Recent reports indicate that al-Qaeda leader Osama bin Laden may have been in contact with LeT leaders until shortly before his death (Parashar 2012). However, LeT appears to currently have an antagonistic relationship with al-Qaeda.
- *Jaish-e-Mohammed.* JeM is an Islamic militant group in Pakistan that was initially founded to counter LeT's growing influence. Although rivals, at the height of violence in Kashmir the two organizations are believed to work together within Jammu and Kashmir.
- *Hizb-ul-Mujahideen.* LeT is believed to regularly carry out joint operations with HuM in Jammu and Kashmir.
- *Taliban.* As LeT shifts its focus from the Kashmir conflict to Afghanistan, it is establishing links with the Taliban to cooperate in confronting international forces there (Times of India 2010).

The constant interactions between LeT and other Islamist jihadist organizations allows LeT to carry out operations in areas where they need expertise. Organizations share logistical support networks, which include local guides, safe houses, and financial support networks. By eliminating links between LeT and such organizations, the easy flow of material, personnel, and other resources to LeT can be reduced and hamper the execution of complex terrorist operations planned by LeT.

Policy P1 also requires the Pakistani government to still maintain a government ban on LeT. On several occasions, pressure by the Pakistani government (which includes various kinds of restrictions on LeT activity) has reduced violent LeT operations.

Every terrorist organization has finite resources in terms of personnel, material, and money. Policy P1 places LeT under considerable stress because it:

- Strains the organization financially (through government bans) forcing the organization to devote some resources to circumventing sanctions;
- Strains the organization internally in order to handle intra-organizational conflict and splintering within the organization—the effort spent on planning terrorist attacks is reduced by the amount of time the organization needs to deal with internal conflicts;

- Reduces the resources available to the organization by cutting its ability to tap other Islamist organizations for resources;
- Hampers the organization's operational capabilities by disrupting its training camps and its military support—LeT would need to use some resources to focus on building training facilities and replacing its military support with other sources instead of using those resources to carry out terrorist attacks;
- Reduces the flow of recruits to LeT and reduces their support among the general population (from which recruits are drawn) by targeting LeT's communications capabilities and infrastructure;
- Reduces the rationale offered by LeT to justify its terror attacks by reducing the creation of martyrs (or *shahids*) who are essential to LeT's recruiting effort. Martyrs are frequently cited in LeT publications as worthy examples for young men to emulate.

Elements of policy P1 have certainly been tried before, though it does not appear that the entirety of policy P1 has been implemented previously. The discussion belows considers only two implementing parties—the US and India.

Disrupting the internal cohesion of LeT means taking actions to create additional internal friction between LeT members and increasing mutual suspicion between LeT operatives. This does not necessarily apply only to the upper echelons of LeT—sowing mutual suspicion amongst lower level elements can also be beneficial as it has the potential to reduce LeT's tactical operations. The danger inherent in the strategy of disrupting LeT's internal cohesion at the upper echelons is that the current LeT leadership might be replaced by even more militant lower level commanders who significantly increase LeT's terrorist activities. This strategy should ensure that the most militant LeT operatives are targeted rather than just the leaders. Exactly who to target would need to be based on detailed intelligence about the internal functioning of LeT that can be obtained via phone and email surveillance, as well as human intelligence (when available). However, LeT considers its dual missions of *jihad* and *dawa* of equal importance and thus has a large infrastructure providing education, outreach, medical care, and propaganda. Leaders from those sectors of the organization may be less inclined to violence, especially when the consequences threaten their own work.

However, disrupting LeT's support from the Pakistani military might be perceived by Pakistan as an act of war. It has long been believed (Lakshmi 2012) that Pakistan's military keeps anti-India sentiment stoked to a high level to justify the importance of the military within Pakistan. Any overt attempt to disrupt Pakistani military support for LeT by India may be viewed by the Pakistani military as a hostile act to which they are obliged to respond. Thus, there are some risks to India in pursuing this policy. Likewise, there are risks to the US in pursuing such an action. US forces in Afghanistan currently depend on Pakistan for providing supply routes, which Pakistan has shut down during periods of tension. Nonetheless the US and India should explore sanctions and other forms of diplomatic pressure to induce Pakistan to cease supporting LeT as a proxy force.

Disrupting the communications capability of LeT is also difficult. Tactical methods to disrupt such communications run the risk of discovery if these operations are carried out on site (e.g., if US or Indian operatives are discovered carrying out these operations), invoking the spectre of an angry Pakistani military using such discoveries to stoke anti-US and anti-Indian sentiment within the Pakistani population. The Raymond Davis affair, in which a CIA contractor killed a pair of Pakistani intelligence operatives in 2011 led to a major crisis in Pakistani-American relations and gives a sense of the possible consequences of US operations within Pakistan (BBC News 2011). However, offensive cyber-operations that can be carried out from locations outside Pakistan's borders run a lower risk. Such a strategy would combine technical means to hamper LeT's ability to disseminate its message with information operations that both distorted this message and challenged it. However on-ground operations should also be considered.

As maintaining sustained international pressure on Pakistan not to release LeT operatives from jail does not seem to have any obvious drawbacks, it may also be considered.

The strategy of targeting LeT training camps carried greater risks and traditional kinetic operations may also be viewed by the Pakistani military as an attack on Pakistan. As with undermining LeT's communcations capabilities, a combination of technical means and information operations could be used against the training camps. Training camps require electricity, water, communications, and supplies—any of which can conceivably be disrupted by remote operations.[1] These technical steps could be combined with information campaigns that publicize mistreatment and corruption at training camps. A study of individuals who left terrorist groups found that difficult conditions at training camps was one reason why many recruits abandoned terrorism (Jacobson 2010). Last but not least, it is possible to disrupt training camps through covert cyber operations aimed at convincing the individuals there that hostile forces are closely monitoring them.

Disrupting LeT's relationship with other jihadist players also poses a challenge, but one where cyber operations can again play an important role. Suspicions already exist between operatives in JeM and HM, both of which periodically collaborate with LeT. Stoking these suspicions through forged emails and instant messages or by leaking embarrassing internal communications via well-designed counter-terror cyber operations may lead to the desired disruption. With high plausible deniability, these operations have the potential to complicate LeT operations while offering relatively low risk.

Last, but not least, requiring a Pakistani government ban on LeT through sustained international pressure does not seem to pose any obvious risks either to the US or to India.

---

[1] For example, Pakistan's electricity system is already strained, so further disruption will not be a great challenge. Much of Pakistan's residential water system relies on tube wells (Haydar et al. 2009). Remote controlled devices either in pipes or the wells themselves could be used to make water access from LeT camps unreliable and complicate LeT operations.

11.2 Policy P2

Policy P2 differs from Policy P1 in just one respect. Like P1, it too requires a government ban on LeT. However, instead of requiring that the relationship between LeT and other militant Islamist organizations be disrupted, it recommends that overseas trials of LeT members be carried out "under wraps".

The data collected and analysis specifically referred to trials of LeT members in Australia. A case in point is that of Faheem Lodhi (also known as Abu Hamza), a Pakistani-born Australian citizen, who was arrested in October 2003 for plotting kinetic attacks against Australia's power grid. Other potential targets included various military facilities in and around Sydney. Lodhi was convicted to 20 years in jail in August 2006, with no possibility of parole for 15 of those 20 years. Lodhi's case was widely covered in both the Australian and Pakistani press.

Lodhi's operation was closely linked to that of Willie Brigitte. A French-Caribbean convert to Islam, Brigitte was suspected of having connections not only to LeT, but also to al-Qaeda. After attending training at LeT camps in Pakistan, Brigitte is believed to have come under the direction of LeT commander Sajid Mir. Mir, in turn is alleged to have coordinated LeT's foreign recruits and is also believed to have played a major role in the 2008 Mumbai attacks (Rotella 2011). There is substantial evidence suggesting that Mir is part of Pakistan's Inter Services Intelligence agency. Brigitte bought a one-way ticket from France to Australia, married an Australian woman who had served in the military, and scouted out possible military targets in Australia. France's intelligence agency, Direction de la Surveillance du Territoire (DST) shared information about Brigitte with the Australian Security Intelligence Organization (ASIO) which arrested Brigitte in October 2003. He was deported to France shortly thereafter. Brigitte was tried in February 2007 and convicted in March 2007. His case was also widely covered by the press both in France and Pakistan.

Relative to its size, Australia has had a number of high-profile trials of LeT operatives. While the system did not generate rules on the effects of international trials overall, a hand-count indicates that in the 25 months in which an international trial (a trial outside of south Asia—including the United States and France) of LeT personnel took place, armed clashes between LeT and Indian security forces occurred in 15 of them, strongly indicating that international trials, at best, serve little deterrant value against LeT.

The data and rules show that the wide publicity generated by Lodhi's trial was *not* helpful to the cause of reducing LeT terror attacks. The year 2006, when there was wide coverage of Lodhi's trial in Australia and Pakistan, also saw the largest number of attacks planned by LeT as shown in Fig. 11.2 (we caution the reader that we are not implying that the attacks were caused by the trial) (Fig. 11.2).

The data, TP-rules, and Policy Computation Algorithms do not explain *why* LeT launches attacks after trials of LeT terrorists. It is possible that high-profile international trials become causes for LeT that can be used to motivate the rank and file LeT members.

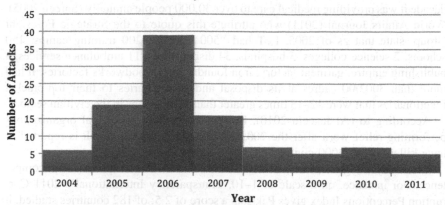

Fig. 11.2 LeT attacks from 2004 to 2011

The above evidence suggests that such trials should be carried out in appropriate venues and under appropriate legal safeguards, but with limited reporting allowed by the press and with the barest minimum of public statements from law enforcement and politicians. Where possible, courts should issue appropriate gag orders to the extent permitted by local and national laws.

As Policy P2 is almost identical to Policy P1 except for two things—disrupting LeT support to other Islamist organizations, and cutting down the publicity generated by foreign trials—the cons of adopting Policy P2 are almost identical to those for policy P1. Disrupting LeT support for Islamist organizations can be done in a manner similar to that for Policy P1. However, cutting down the publicity generated by foreign trials of LeT personnel runs the risk that these trials will be dismissed as a sham, without the usual transparency expected of courts in the West. This, in turn, can be used to stoke anti-Western sentiment in Pakistan, enabling the Pakistani military to justify the sense of paranoia in Pakistan. As a consequence, even though publicity for such trials should be limited, extreme effort must be taken in order to ensure that the fairness and integrity of the process is transparent.

11.3 Policy P3

Policy P3, like Policy P1 above, requires that the relationship between LeT and other militant Islamist organizations be disrupted. However, in addition, it requires undermining LeT social service programs.

Much like Hezbollah in Lebanon and Hamas on the West Bank, there are areas of Pakistan where LeT is woven fully into the fabric of society through the provision of a wide array of social services. LeT runs numerous schools, ambulances, hospitals, and clinics throughout the country.

In the early 1990s, LeT began establishing a social welfare wing. Within a decade it was providing medical care to over 30,000 people annually (Sareen 2005). World Affairs Journal (2011) who attribute this quote to the Strategic Foresight Group, state that as of 2005, LeT had "500 offices, 2200 training camps, 150 schools, 2 science colleges, 3 hospitals, 34 dispensaries, 11 ambulance services, a publishing empire, garment factory, iron foundry, and woodworks factories. It had more than 300,000 cadres at its disposal and paid salaries to their top-bracket functionaries that were 12–15 times greater than similar jobs in the civilian sector."

According to (Al Jazeera 2010), LeT was one of the principal organizations performing relief work after the 2005 Kashmir earthquake, which is reported to have killed over 75,000 civilians.

The Pakistani government is often accused of both corruption and incompetence. For instance, on a scale of 1–10, Transparency International's 2011 Corruption Perceptions Index gives Pakistan a score of 2.5; of 182 countries studied, it is ranked 134 (higher ranks denote higher levels of corruption). (Hussain 2011) provides an analysis accusing the Pakistani government of being both incompetent and corrupt. Its ability to offer basic social services to its citizens is highly questionable. When organizations like LeT provide basic services to Pakistan's citizens, those citizens develop a strong bond with the organization.

Few would argue against providing basic services to the many good people of Pakistan. But it is imperative that LeT's growing influence be checked, thereby cutting the steady stream of new recruits and supporters to LeT.

Policy P3 differs from the previous policies in one action, namely disrupting the social services provided by LeT. LeT provides valuable social and medical services throughout Pakistan and these can be used to justify their importance to the ordinary Pakistani citizen.

One cause for this is the failure of the Pakistani state to do so, leading to a vacuum that LeT stepped into fill. Disrupting LeT's social services may lead to a backlash as ordinary citizens who relied on LeT's ambulances and doctors suddenly find themselves without these services. The best option is to offer well-branded competing services through appropriate aid agencies or international NGOs like the Red Cross or Medecins sans Frontiers as well as to press the Pakistani government to meet the needs of its own people.

Also, LeT is committed to its dual missions of *dawa* and *jihad*. If its social services (which are crucial to *dawa*) are complicated by sanctions due to terrorist activity, the organizations will be forced to make choices about which activity is its true priority.

11.4 Policy P4

Policy P4 is a combination of Policies P2 and P3, requiring that countries interested in disrupting LeT's operations should both undermine LeT's social service programs, as well as reduce the publicity generated by high-profile trials of LeT personnel in foreign countries.

As all the individual actions in policy P4 already occur, in various combinations, in the preceding policies, the pros and cons of these actions are identical to those discussed before.

11.5 Tactics: Increased Non-Violent Covert Action Against LeT

All policies generated by the Policy Computation Algorithm recommend that:

- Splintering of LeT should be encouraged;
- Intra-organizational conflict within LeT should be encouraged;
- Efforts should be made to cause resignations by LeT leaders—possibly through sustained pressure by the international community to force Pakistan to clamp down on LeT;
- The relationship between LeT and the Pakistani military should be disrupted;
- LeT's public communications capabilities should be disrupted;
- LeT's training camps should be disrupted.

Non-violent covert action is one mechanism to achieve some of these strategic policy objectives. There are some precedents for eliminating terrorist groups by exacerbating their internal tensions and playing them off against other groups.

Historically, organizing terrorism involves the complex task of balancing the priorities of security, efficiency, and control. Because terrorist organizations are clandestine, communications (of any form whether it is electronic or personal meetings) must be limited because they can lead to detection by security forces (Lindelauf 2011). This limits the ability of leaders to oversee operations. This absence can lead to inefficiency, the misuse of organizational resources, or lack of control that might result in terrorist operations that are counter-productive for the organization. However, the only solution to these problems is more communications, which increases the organization's vulnerability (Shapiro 2007).

Most famously, the Abu Nidal Organization—perhaps the world's most notorious terrorist group in the 1980s—was effectively defanged by intelligence operations that persuaded Abu Nidal that his top aides were betraying him. Crucial to this operation was intelligence from another terrorist organization, the PLO, which had a bloody rivalry with Abu Nidal (Perry 1992). Abu Nidal's crucial vulnerability was his personal paranoia. LeT is much larger and has more public operations, which could become a different sort of weakness. Highlighting personal rivalries and corruption within LeT ranks could undermine organizational cohesion and lead to more infighting. Addressing these problems could be an irritant to the organization's leadership.

LeT faces a particular problem in that its terrorist operations could interfere with its missionary work. Violence could be counter-productive if it leads to crackdowns on LeT and the social services it provides. There are a range of technical and psychological information operations that could be employed to complicate LeT operations and press it to choose between its two priorities, *dawa* and *jihad*.

However, one possible consequence of terrorist groups seeing a decline in control is greater violence by those most committed to terror. In reviewing Headley's testimony, Tankel comes to the conclusion that this may have been the case in the 2008 Mumbai attack, which was originally planned as a much smaller operation. The operation became larger as LeT operatives disillusioned with their organization's sitting on the sidelines in Pakistan's conflict sought a new outlet for violence (Tankel 2011a, b). Nonetheless, the model presented shows that using covert action and information operations to sow dissent in LeT could help to induce the organization to move away from violent conflict. Further, if LeT as a whole were moving away from violence, then the Pakistani state, out of self-interest, might be more amenable to cracking down on violent splinter groups because these groups would be less subject to ISI control and could turn their animus against the state.

There are many forms such covert action could take:

- Covert action could include explicit actions to sow mistrust amongst insurgent commanders (e.g., by leaking credible reports of collaboration by some leaders with forces arrayed against LeT, or by leaking credible reports about inter-organizational rivalry). Such credible reports could include falsified communications. As mentioned earlier, such covert actions should target LeT operatives who are deemed to be the "worst" so that a "more reasonable" operative is not replaced by a "more militant" operative, thus worsening an already bad situation.
- Targeting specific LeT commanders that leave credible evidence implicating another leader within LeT, e.g., via falsified emails and phone records. In order to avoid the possibility that an LeT commander is replaced by a more violent commander, such operations should choose targets with extreme care.
- Complicating LeT operations by targeting facilities through unpredictably disrupting electricity, water, and phone services leading into the facilities and/or sporadically jamming wireless frequencies within a narrow area in and immediately around such facilities. The technical capabilities to do so almost certainly exist. In Operation Orchard, Israeli operatives allegedly infiltrated the network environment around Syrian radar and installed a "kill" switch that subsequently allowed them to carry out a covert air strike against a Syrian nuclear facility (Adee 2008). Targeting the information systems of LeT and its allies, both to hamper operations and to gain valuable insight into the organization's operations should be a relatively simple endeavor by comparison. The information gained through such operations could be used to better understand internal organizational dynamics and better target information campaigns. Finally, if LeT cannot trust its own communication networks, it would be forced to identify more expensive, lower-tech communications methods, which would further complicate operations.

Table 11.3 Probability that disrupting LeT's internal cohesion would largely eliminate many different types of attacks in a 1–3 month time frame after LeT's internal cohesion was disrupted

Type of attack	Probability that disrupting internal cohesion will significantly reduce attacks within 1–3 months (%)
Attacks on symbolic/tourist sites	75
Attacks on public sites	87.5
Attacks on professional security forces	37.5
Attacks on security installations	50
Attacks on civilians	100
Attempted attacks	62.5
Attacks on transportation	100
Attacks on the Government	87.5
Attacks on holidays	87.5
Armed Clashes	37.5

Intelligence officers with experience in planning such covert operations will think of many other ways of disrupting the smooth operations of a terror group. The goal of this book is to provide *strategic* recommendations that can be implemented via a variety of tactical means on the ground by seasoned intelligence operations experts. These suggestions are not intended to be comprehensive, but rather are broad proposals on how to undermine LeT's internal cohesion.

The more interesting question derived from the data was: how effective would covert action to disrupt LeT's smooth functioning be in terms of reducing the number of attacks carried out by LeT? Analyzing the data set to understand this question, resulted in the findings seen in Table 11.3 above.

Table 11.3 shows that finding ways to disrupt the internal cohesion of LeT provides a very strong method to significantly reduce many kinds of attacks—but not all. As Table 11.3 indicates, attacks against professional security forces, security installations, and armed clashes with security forces are still likely to occur. Most such LeT attacks target India.

Table 11.3 also illustrates why merely disrupting the internal cohesion of LeT is not enough to reduce the intensity of different types of attacks. It explains why each of the policies we have suggested include many types of actions besides disrupting internal cohesion of LeT—these other actions help reduce attacks on professional security forces, security installations, and armed clashes.

It is also noteworthy that even though the Policy Computation Algorithm eliminated some rules dealing with attacks on holidays because of infeasibility, it is clear from Table 11.3 that methods that disrupt LeT's internal cohesion also have the side effect of having a reasonably high chance (87.5 %) of eliminating attacks on holidays as well.

Nonetheless, while targeting LeT's internal cohesion, it is important that Indian security forces remain vigilant, as they are highly likely to continue being targeted by LeT.

11.6 Tactics: Deterrence and Coercive Diplomacy

Every policy generated by the Policy Computation Algorithm requires the disruption of the relationship between LeT and the Pakistani military.

Disrupting the relationship between LeT and the Pakistani military has been difficult to achieve via diplomatic means. After a major attack (such as the Mumbai attacks in which LeT was decisively shown to be the perpetrator—and where strong evidence points to the involvement of Pakistan's Inter Services Intelligence agency), Pakistan typically bows to international pressure, arresting and imprisoning LeT leaders along with large numbers of the rank and file. A few months later, they are usually released. Nonetheless, the periods of relative quiet following major crackdowns indicate that the Pakistani state is capable of reducing and controlling LeT operations (Tankel 2011a, b).

But, until there are hard consequences for the Pakistani government as a result of its support for LeT, it seems unlikely that they will discontinue their relationship.

Ideally, the United States and India could employ coercive diplomacy in which Pakistan is punished through non-violent means for its support of terrorist organizations. Unlike deterrence, in which one actor seeks to persuade an opponent that the costs of a particular action will outweigh the benefits, coercive diplomacy uses threats or limited use of force to persuade an opponent to undertake an action, such as withdrawing from captured territory (Craig and George 1990). In the case of LeT, this was first suggested—in the context of a game theoretic study (Dickerson et al. 2011)—by three of the authors of this book. There are many ways of applying coercive diplomacy—a few such ways are listed below. However, as noted by Nobel Laureate Tom Schelling throughout his celebrated book on *The Strategy of Conflict*, it is critical that any threats be followed up by real action should the circumstances that trigger the threat actually come to pass (Schelling 1980). In other words, if a party did threaten to take one of the actions listed below, it must do so publicly (so the Pakistanis know that the issuer of the threat, e.g., the US administration, has to follow through on the threat). If a determination is made that LeT did participate in some kind of major terror operation, then the party issuing the threat must follow through on the threat—otherwise the threat is not a credible future deterrent.

However, finding methods to compel the Pakistani government to change its behavior has been difficult. Because Pakistan is a nuclear power, threats to use force against it risk escalation. In 1999, after Pakistan seized the Kargil region of Kashmir, Indian forces re-took the area (Swami 2007). But after the 2001 Parliament attack,[2] India mobilized forces along Pakistan's border but found itself with no options in the face of a Pakistani readiness to consider the use of nuclear weapons (Coll 2006). For the United States, Pakistan has played a crucial role in fighting al-Qaeda and as the critical supply route for NATO forces in Afghanistan.

---

[2] There are differing opinions about LeT's role in this attack as JeM operatives were arrested and tried in connection with the attack.

There are modest prospects that non-violent options to compel Pakistan to change its behavior will improve. First, the United States is reducing its presence in Afghanistan, which will also shrink Pakistan's leverage due to control of critical supply routes. Second, Pakistan is facing multiple internal crises that threaten the very coherence of the state. There are multiple internal violent conflicts with sectarian, ethnic, and political origins against a backdrop of a continuing economic crisis, a decaying infrastructure, and a fast-growing population (Rashid 2012). These difficulties may make Pakistan more amenable to pressures and incentives. A few possibilities are discussed below:

- *Tying foreign aid to reduction of support to LeT.* Pakistan can be told by leading donors (e.g., the US) that military and development aid might be significantly reduced if LeT carries out a certain level of terrorist acts in the future. Leading donors can also influence international multilateral banks like the World Bank, the IMF, and the Asian Development Bank to adopt such policies. However, to be effective such threats must be credible.
- *Water Rights and Coercive Diplomacy.* Another non-violent form of sanctions could include restricting water flow into Pakistan. For instance, if Pakistan is warned that a certain percentage of Indus water would not be allowed to flow into Pakistan if LeT carries out certain attacks, then this could serve as an inhibitor to such attacks. The headwaters of the Indus River have their source in India (Kugelman and Hathaway 2009)—though the Indus Waters Treaty between India and Pakistan is over 50 years old and has never been broken, it could be used as a credible threat to ensure better behavior from the Pakistani government. Needless to say, threats to use water as a counterterrorism tool must be weighed with extreme care—no one wants to affect innocent civilians. The point we make is that a credible threat to dial down water supplies if LeT carries out future terrorist actions may pose a deterrent to such future attacks. However, this again has risks. Pakistan's military may respond by increasing its anti-India rhetoric and stoking fears of Indian action against Pakistan. But they might also consider it an act of war. The point here is to tie threats to dial down water supply directly to clear terrorist attacks orchestrated by LeT or affiliated groups.
- *Trade Sanctions.* Pakistan derives substantial revenues from the textile business—in fact, textiles account for approximately two-thirds of all its export revenues and accounts for nearly 40 % of the formally employed labor force (World Bank 2006). Any credible threat to impose restrictions on textile imports from Pakistan should there be a Pakistani hand in LeT-backed terror attacks may present a powerful deterrent to terrorist activity by Pakistan.
- *Limited Military Action.* Although Pakistan is a nuclear power and taking military action against it exceptionally risky, there are some precedents for limited military responses against nuclear-armed states. We do not recommend military action be taken—but the military option must remain on the table.

Note that in all of these cases (except for the last one), a *non-violent threat* is made, primarily as a deterrent. The threat is contingent upon a harmful action carried out by LeT or one of its proxies. The threat must be made publicly with clear-cut conditions for what retaliatory actions will be taken in the event that an LeT-backed terror attack (including attacks by its proxies) occurs. The *public* aspect of the threat makes it difficult for the threatening party not to follow through on the threat should an LeT-backed attack occur. When LeT and the Pakistani government both know that is the case, and as they believe the threatening party would be obliged to take the punitive steps in the event of an LeT attack, the Pakistani government has a strong incentive to pressure LeT to behave more responsibly.

11.7 Tactics: Increased Non-Violent Covert Action Against LeT Communications Campaigns

LeT carries out communications campaigns propagating their worldview. As discussed above, LeT values jihad (holy war) and dawa (preaching) equally and believes the two missions support one another. LeT has built an extensive communications infrastructure including magazines, rallies, and an internet presence to advance its worldview (Rana 2006).

Previous chapters discussed the existence of a strong relationship between LeT's ramping up a publicity campaign and a variety of attacks.

- Chapter 6 showed how communications campaigns by LeT are strongly correlated with attacks on professional security forces in the next 1–3 months.
- Chapter 7 showed how communications campaigns by LeT are strongly correlated with attacks security installations in the next 1–3 months.

Both of these chapters indicate that there is a strong relationship between LeT's use of the press and communications media, and subsequent attacks on security forces and installations—usually in India. As in the case of Sect. 10.2, these two kinds of attacks do not appear to be easily reduced by disrupting LeT's internal cohesion. However, it is possible that taking steps to diminish LeT's use of the press, news, and periodicals media can impact their predilection to launch attacks on professional security forces and security installations. On a broader basis, permitting LeT to propagate its message of jihad further radicalizes Pakistan's population, which increases the number of recruits to radical organizations and decreases the possibility of more peaceful relations between India (or the US) and Pakistan.

Disrupting LeT's relationships with the news media and periodicals is particularly complex because there is a distinction between undermining the message LeT sends and disrupting the mechanism by which it is sent. In a typical week, LeT distributes tens of thousands of copies of periodicals throughout the country and has officials delivering sermons in mosques nation-wide. Similar to the discussion above on fomenting internal dissent, disrupting LeT's communications capability will require a combination of technical means and counter-messaging. A few possible mechanisms are described here:

- *Disrupting LeT's ability to communicate with the media involved in news/ periodicals.* As mentioned above, simple cyber-attack techniques can be used to hack LeT's computer systems, diverting emails directed at news media and periodicals to "nowhere" (e.g., by sending a "bounced email" message). A better option forwards the message to the intended recipient as well as to a designated location where the anti-LeT organization can read the message and plan how best to counter LeT's message before it has been delivered, while LeT continues under the impression that its operation has not been compromised. Simple hacking and communication techniques can also be used to direct all phone calls emanating from fixed line and mobile phones located in areas designated "LeT locations" to a third party that pretends they are the legitimate recipients of LeT phone calls (or simply allows a third party to listen into the call). A more sophisticated variation might be to issue communications in LeT's name that embarrasses the organization and undermines its message.
- *Disrupting the "airwaves" between LeT and a news outlet or a periodical publisher.* Counterterrorist organizations could disrupt the spectrum usage (including mobile phones) in and around LeT locations and training camps via appropriate, hard to pin down, diversionary and/or disruptive tactics including hacking the communications between phones and proximate cell towers. Any phone that is frequently within the region of a cell tower that intersects with a region designated an "LeT region" can be labeled a "suspect phone". If communications with all phone towers are hacked, such phones can be identified whenever they are turned on (as they need to be recognized by a cell tower). There is some precedent for employing this strategy against terrorist groups. According to (Reyes and Dudley 2006), beginning in 2003, the United States Drug Enforcement Agency infiltrated satellite phones embedded with listening devices into the FARC in Colombia. This infiltration was reportedly critical to the string of counter-terror successes against the FARC later in the decade. Such tactics would need to be covert (unlike the case of Colombia) as the Pakistani military is not likely to be supportive. These techniques can be used to gather intelligence and complicate LeT's communications. Even if these methods are detected, they will force LeT to adopt expensive counter-measures to ensure their security.
- *Disrupting a news media/publisher's site.* This is not a preferred option, but is discussed here for the sake of completeness. Clearly, most publishers of material supportive of LeT are well known and easy to identify. Using covert methods as discussed earlier, their electricity, water, Internet, and communications media can be disrupted. A weaker form of disruption might be merely to monitor their phone calls for calls with "suspect phones" as defined in the previous bullet. When such a phone call occurs, the hacked routing system of the cell phone infrastructure can direct the call elsewhere (or do so randomly, not always to induce doubt on the part of the individuals involved). Or, as mentioned above, the call can be allowed to proceed, but everything will be recorded electronically, providing valuable intelligence to any third party interested in disrupting LeT's violent actions.

- *Effective Public Diplomacy.* Attacking LeT's public relations machinery is a useful step, but no substitute for an effective public diplomacy campaign that neutralizes LeT's message by countering their narrative and encouraging alternative moderate voices. Advances in communications technology and the growing ubiquity of mobile phones have led to a vast range of new methods for delivering these messages. LeT and other radical Islamists have made creative use of them, and it is essential that those who oppose these organizations do the same.

While LeT has been sophisticated in its use of technology (Stern 2003), Pakistan overall is a relatively low-tech society. Mobile phones are ubiquitous, but internet access (and even literacy) are at fairly low levels—particularly in rural areas and among the less affluent parts of the population. The challenges of public diplomacy are beyond this monograph. But the importance of LeT's communications infrastructure to its mission indicates that leaving the national discourse in the hands of LeT and its ilk will guarantee generations of violence.

11.8 Conclusion

The statistician, George Box, famously remarked, "essentially, all models are wrong, but some are useful" (Box and Draper 1987, p. 424). The one presented here is no exception.

In this chapter, we have shown that the Policy Computation Algorithm automatically generated 8 policies from our Lashkar-e-Taiba data set—many of these policies are similar to one another and vary only in very subtle ways. As a consequence, we chose to present 4 policies in this chapter.

The policies themselves are significant in the degree of agreement amongst them. All policies agree on twelve things (seven actions to be performed, five actions to be avoided). Specific high points include:

- Targeting the internal cohesion of LeT;
- Targeting the support provided by the Pakistani military to LeT;
- Targeting the communications campaigns launched by LeT;
- Targeting the training bases from which LeT operates;
- Encouraging the resignations of senior LeT leaders;
- Stopping the release of arrested LeT personnel;
- Not engaging in campaigns to arrest or kill LeT personnel, or issuing arrest warrants for them, and not raiding LeT establishments[3];

---

[3] We emphasize that this does not mean dropping law enforcement activity against LeT, just that targeted efforts to arrest/kill LeT members and/or ban the organization have had mixed success and may not be having the desired effect. That said, LeT members can certainly be arrested/killed as part of ongoing operations. No one would argue against the arrest, for instance, of Ajmal Kasab, the lone surviving bomber in the 2008 Mumbai terror attacks.

- Not making explicit efforts to encourage defections by low level LeT personnel.

In addition to agreement on all of the above parameters, the four policies we study suggest

- Targeting social services and medical programs run by LeT;
- Reduce the publicity generated by trials of LeT members, especially those in Australia;
- Ensure the existence of a Pakistani government ban on LeT and
- Target the relationship between LeT and other Islamist organizations.

However, these policies do not offer a silver bullet to reducing the terrorist attacks carried out by LeT. As LeT and its sponsors try to achieve their strategic objectives, they will try to continuously adapt to the counter-terror efforts used to target them.

As a consequence, our policies against LeT must undergo constant change, adapting to behavioral changes in the organization in near real-time. Fortunately, the computational technology introduced in Chap. 3 to automatically learn TP-rules describing the behavior of LeT, and the policy analytics discussed in detail in Chap. 10 to generate the policies discussed in this chapter, do not need to be re-implemented. As new data on LeT's behavior becomes available, these tools can be used to continuously suggest new policy options to policy makers who can then adopt, modify, or reject them as they deem fit.

References

Adee, S. (2008). The hunt for the kill switch. *IEEE Spectrum, 45*(5), 34–39.

Al Jazeera (2010). *Jamaat chief rejects Indian charges, 18 Feb 2010* http://www.aljazeera.com/news/asia/2010/02/201021785121810598.html

BBC News (2011). *CIA contractor Ray Davis freed over Pakistan killings, 16 March 2011* http://www.bbc.co.uk/news/world-south-asia-12757244

Box, G. E. P., & Draper, N. R. (1987). *Empirical model-building and response surfaces.* New York: Wiley.

Coll, S. (2006). The stand-off. *The New Yorker, 13 Feb 2006* http://www.newyorker.com/archive/2006/02/13/060213fa_fact_coll

Craig G., & George, A. (1990). *Force and statecraft: Diplomatic problems of our time* (2nd ed.). New York: Oxford University Press

Dickerson, J, Mannes, A, Subrahmanian, V. S. (2011). Dealing with Lashkar-e-Taiba: A multi-player game-theoretic perspective. In *Proceedings of the IEEE international symposium on open-source intelligence and web mining* (pp. 354–359), Athens.

Ganor, B. (2005). *The counter-terrorism puzzle: A guide for decision makers.* New Brunswick: Transaction Publishers.

Haydar, S., Arshad, M., & Aziz, J. A. (2009). Evaluation of drinking water quality in urban areas of Pakistan: A case study of southern Lahore Pakistan. *Journal of Engineering and Applied Science, 5,* 16–23.

Hussain, B. (2011). *If Pakistan's leaders cannot be honest, at least let them be competent.* UK: The Guardian 2011.http://www.guardian.co.uk/commentisfree/2011/apr/09/pakistan-competent-honest-corruption

Jacobson M. (2010). *Terrorist Dropouts: Learning from Those Who Have Left*. Washington Institute for Near East Policy Policy Focus 101 http://www.washingtoninstitute.org/policy-analysis/view/terrorist-dropouts-learning-from-those-who-have-left

Kugelman, M., & Hathaway, R. (2009). *Running on empty: Pakistan's water crisis*. Woodrow Wilson Center report http://www.wilsoncenter.org/publication/running-empty-pakistans-water-crisis

Lakshmi, R., & Leiby, R. (2012). *India, Pakistan leaders pledge improved relations, The Washington post, 8 April 2012* http://www.washingtonpost.com/world/india-pakistan-leaders-pledge-improvedrelations/2012/04/08/gIQAxI9Q3S_story.html

Lindelauf, R. (2011) *Design and analysis of covert networks, affiliations, and projects*. Ph.D. Thesis, Tilborg University, The Netherlands.

Parashar, S. (2012). Hafiz Saeed's brother-in-law Abdul Rehman Makki is a conduit between Lashkar-e-Taiba and Taliban. *Times of India, 5 April 2012* http://timesofindia.indiatimes.com/world/pakistan/Hafiz-Saeeds-brother-in-law-Abdul-Rehman-Makki-is-a-conduit-between-Lashkar-e-Taiba-and-Taliban/articleshow/12539443.cms

Perry, M. (1992). *Eclipse: The last days of the CIA*. New York: Morrow.

Rana, M. A. (2006). *A to Z of Jihadi organizations in Pakistan (Translated by Saba Ansan)*. Pakistan: Mashal Books http://www.desistore.com/jehadiorg.html

Rashid, A. (2012). *Pakistan on the brink: The future of America, Pakistan, and Afghanistan*. New York: Viking Press.

Reyes, G., & Dudley, S. (2006) Ex-con helps U.S. deliver satellite phones to FARC. Miami Herald, 10 May 2006.

Rotella, S. (2011). *Pakistan's terror connections: Chicago terrorism tria, what we learned and what we didn't about Pakistan's terror connections*. Propublica. http://www.propublica.org/article/chicago-terrorism-trial-what-we-learned-and-didnt

Sareen S. (2005). *The Jihad Factory: Pakistan's Islamic Revolution in the Making*. New Delhi: Observer Research Foundation

Schelling, T. C. (1980). *The strategy of conflict*. Cambridge: Harvard University of Press.

Shapiro, J. (2007). *The terrorist's challenge: Security, efficiency, control*. Center for International Security and Cooperation, Stanford University http://igcc3.ucsd.edu/pdf/Shapiro.pdf Accessed 27 March 2012.

Stern, J. (2003). *Terror in the name of God: Why religious militants kill*. New York: HarperCollins.

Swami, P. (2007). *India, Pakistan and the secret jihad: The covert war in kashmir, 1947–2004*. Abindon: Routledge.

Tankel, S. (2011). *Lashkar-e-Taiba: Past operations and future prospects, national security studies program policy paper*. New America Foundation, Washington.

Tankel, S. (2011b). *Storming the world stage: The story of lashkar-e-taiba*. London: C. Hurst & Co.

Times of India (2010) *Lashkar main Taliban ally in Afghanistan: US forces, 12 July 2010* http://articles.timesofindia.indiatimes.com/2010-07-12/india/28297502_1_taliban-commander-taliban-operatives-isaf

Wall Street Journal. (2008) *Lashkar-e-Taiba served as gateway for western converts turning to jihad, 4 Dec 2008*.

World Affairs Journal (2011) The next al qaeda? lashkar-e-taiba and the future of terrorism in South Asia http://www.worldaffairsjournal.org/article/next-al-qaeda-lashkar-e-taiba-and-future-terrorism-south-asia

World Bank (2006) Pakistan growth and export competitiveness, World bank report 34599-PK, April 2006.

Appendix A
Data Methodology

This study used a systematic method to analyze the behavior of LeT, learn a behavioral model of LeT, and make forecasts and predictions about how LeT will behave in the future in both real and hypothetical situations. Many of these forecasts have already proven true. Finally, we have developed principled methods to learn how LeT's terrorist behaviors may be "reined in."

This *appendix* describes the methodology followed in arriving at the results described in this book. The four-step process is explained below.

- Step 1 *Systematically Gather Data*

  Data was gathered by studying a wide variety of open source literature (mostly news articles from press sources worldwide, but also a range of papers, books, and reports published by a range of researchers from academia, NGOs, and government and private research institutes) to gather data. A total of 770 variables that needed to be considered were identified. In general, data sets of this nature focus on a particular aspect of a group's behavior and are shaped by the interests and methods of the particular discipline studying an issue (e.g., databases assembled by economists focus on economic factors) and test particular hypotheses. This study is not constrained by any one discipline and includes a broad range of variables covering economic, social, cultural and organizational factors in the group's behavior.

  The LeT data set includes, whenever possible, a quantitative value for each variable (e.g., number of people killed in fedayeen[1] attacks, or number of actual fedayeen attacks carried out by LeT). The data set captures two sets of variables—*environmental variables* which includes variables describing the organizational structure and behavior of LeT along with aspects (e.g., social,

---

[1] A *fedayeen* attack is one where the attackers are prepared to die. However, unlike suicide attacks, fedayeen attacks do not involve attackers who wish and expect to die, as in the case of bombers who wear suicide vests.

V. S. Subrahmanian et al., *Computational Analysis of Terrorist Groups: Lashkar-e-Taiba*,   177
DOI: 10.1007/978-1-4614-4769-6, © Springer Science+Business Media New York 2013

political, religious, financial, ethnic) of the environment in which LeT is functioning during a given month. Environmental variables also capture, through specific variables, the behaviors of other actors (e.g., the Pakistani government, the Pakistani military, the US, India, etc.) whose acts may impact the behavior of LeT. The second set of variables are *action variables* that describe various aspects of the violent actions taken by LeT during a given month.

The LeT data set captures information on a monthly basis, starting from January 1989 up to December 2010, and the analysis in this book is based on that data. The next section of this chapter describes our data set in considerable detail.

- Step 2 *Learn Behavioral Models*

The second important step in the framework is the use of sophisticated data mining algorithms developed by to automatically learn models of LeT's behavior. While extremely successful companies like Amazon and Google use data mining on a daily basis, both governments and social science researchers have been slow to embrace the power of data mining, instead frequently arguing that reasoning about terrorist groups is too idiosyncratic a task for computational tools.

The model developed is known as *Stochastic Opponent Modeling Agents (SOMA)*. The SOMA Rule Learning Algorithm automatically learns SOMA models of terrorist groups and has already been applied to learn behavioral models of a number of terrorist groups including Hamas and Hezbollah (Khuller et al. 2007; Mannes et al. 2008a, b). The algorithm considers every bad act (e.g., targeting security infrastructure, or targeting civilian infrastructure, or carrying out fedayeen attacks, to name a few) and automatically learns logic conditions on the environmental variables that neatly distinguish between when the group performs the bad act (perhaps at a given level of intensity) from when it does not. These logic conditions then define probabilistic rules that give the probability that LeT will carry out a given bad act (perhaps at a given intensity level) during a given time frame or under some set of real or hypothetical circumstances. SOMA rules are relatively easy "if then" rules that analysts and policy experts can easily understand and often explain.

Section B of this chapter describes what SOMA rules look like, and what kinds of statistical conditions SOMA rules should satisfy in order to be considered "good".

- Step 3 *Make Forecasts*

Once models of the behavior of LeT have been learned, *forecasts* can be made. This book presents a wide variety of forecasts—and also explains forecasts that have previously been made, some of which have already come out to be true.

Forecasts must satisfy many conditions. First and foremost, forecasting requires *consistent, checkable accuracy*. Many people have forecast things that have turned out to be true—what is not always clear when such claims are made is how often they have made forecasts, and how accurate these forecasts were in general. A "one off" forecast is not a reliable way of measuring the ability to tell the future. Second, forecasts must be made in clear English—otherwise, decision makers and policy analysts who may possess a variety of educational backgrounds may not understand them. Forecasting methods that are purely mathematical (e.g., based on clustering algorithms such as those currently used by a variety of services on the Internet today to forecast purchases—e.g., as Amazon does when they make "suggestions" to customers) are hard to explain and may be of limited utilty.

- Step 4 *Create Possible Policies*

The LeT data set consists of 770 variables, and the values of these variables have been collected on a monthly basis from January 1989 to December 2010. It is impossible for any analyst—even a supremely intelligent one—to analyze this data set "manually".

For reference purposes, Microsoft Excel brings up approximately 15 columns on a single screen when launched. In contrast, for an analyst to even *visually* see the LeT data would require a screen capable of accommodating 770 rows, which would require a computer screen about 50 times as wide as the typical screen. Needless to say, analyzing this data visually to identify important correlations is difficult.

Over the years, multiple systems have been developed to automatically generate *policies*. The TOSCA system (Parker et al. 2011) is very fast. An analyst can "feed in" a database (such as our LeT data set) and specify a goal he wants to achieve (e.g., simultaneously reduce fedayeen attacks by LeT to a certain amount or less and reduce attacks on Indian security installations to a certain amount or less). The system interprets this to mean that these two types of behaviors by LeT must be reduced to a more acceptable level without increasing the intensity of other bad acts by LeT. TOSCA tries to find ways in which the environmental variables can be "reset" so that the probability of achieving the user's stated goal is maximized. TOSCA also allows the policy analyst to specify constraints on do's and don't's.

The environmental variables may serve as policy levers that the analyst can try to use to change a group's behavior. Some levers may be impossible to use— and the analyst can specify these via constraints. Other levers may have high costs—and the analyst can specify these too via cost specifications.

Such a tool helps the analyst leverage the enormous power of modern computing technology to analyze very large multidimensional spaces (in this case, a 770-dimensional space) bringing his own knowledge (via his specification of constraints) to bear as an important input to the problem.

The PAGE (Policy Analytics Generation Engine) (Simari and Subrahmanian 2010) system performs a similar task as TOSCA, but uses the SOMA rules generated directly. However, in this analysis, we chose a simpler method based on (Bell94a, Bell94b) described in Chap. 10.

The rest of this chapter discusses the first two these four steps (and their associated software components) in greater detail. The forecasting methodology and the policy generation methods will be described in Chaps. 3 and 10 respectively.

A.1 Systematically Gathering Data

This section provides a detailed overview of the LeT data set, which contains about 770 variables that can apply to any terrorist group, not just LeT. This set of variables has already been used to gather data about several terror groups including Jaish-e-Mohammed, the Student Islamic Movement of India (SIMI), and Indian Mujahideen (IM) in South Asia region, as well as the Forces démocratiques de libération du Rwanda (or FDLR). Moreover, the *Automated Coding Engine* (ACE for short) has been recently developed to gather data on these 770 variables for any group whatsoever by crawling Lexis-Nexis news feeds and automatically extracting the values of these variables from them. ACE is about 75 % accurate—though it needs human intervention to get the answers "fully correct", it yields significant savings in coding time. However, for this study of LeT, all data was collected manually.

A.1.1 Sources

The data set was collected manually by consulting a wide range of sources. Every effort was used to ensure the sources were authoritative. The sources include:

- International news sources
- National, regional, and local news sources
- Books, papers, and reports authored by various researchers
- Reports and transcripts from authoritative sources
- Selected web sites that track terrorism-related information

Every effort was made to ensure that information we gathered was accurate. All data for this project was gathered *manually* from these sources and are deemed correct by the authors.

A.1.2 Data Organization

The LeT data set is organized as a relational database table. The *rows* of this table correspond to months, starting in January 1989, to Dec. 2010. Thus, there is exactly 22 years of data in this data set. The *columns* of this table correspond to individual variables. As mentioned earlier, the variables fall into two categories:

Environmental variables describe LeT's structure, leadership, and communications, along with social, cultural, ethnic, political, and financial aspects of the environment in which LeT was functioning during these 22 years, including actions taken by other actors whose actions might have had an impact on LeT's behavior.

Action variables, which describe the various types of terrorist actions taken by LeT during these 22 years.

A.1.3 Environmental Variables

We studied a total of 570 environmental variables. These variables fell into several broad categories, listed below in alphabetical order.

- **Communication:** This set of variables describes LeT communications, including the nature of its message and the infrastructure LeT used to propagate this message. Specifically, the variables describe the:
  - *Addressee* of various communications—who was the intended audience of various communications, separated out by entity type (e.g., government versus security forces versus international organizations), by region (e.g., global audience versus national audience), ethnicity (were communications directed at specific ethnicities or groups holding specific beliefs).
  - *Medium* of the communications—did LeT use blogs or emails or web sites to get their message out? Did they use books, periodicals and conferences? Did they enlist clerics? Did they use social networking sites or radio or phone to get their message out?
  - *Messaging* in the communications—these variables describe the message in these communciations. Did LeT use these communications to call for violence, recruit for their cause, or call for a change in lifestyle of the population? Did LeT claim of responsibility for attacks, publicly identify enemies, make confessions, or justify violence?

- **Environment:**
  - *Relationship with Security Forces*—these describe the behavior of security forces in areas where LeT is active, specifically Jammu & Kashmir, Pakistan, Afghaistan, and the rest of India. There are variables for the following actions:

    Bombardment
    Burning settlements
    Corruption
    Imposing curfews
    DDR (demobilization, disarmament and reintegration) programs
    Desertion among security forces
    Executions (of offenders)
    Imposing restrictions on freedom of restriction
    Shutting down public sites
    Sealing off regions
    Paying soldiers on a regular basis

Engaged in repression against civilians
Engaged in sexual violence against civilians
Providing social services
Suppression of the opposition
Engaged in torture

- *Government's International Relations*—this set of variables captures the types of international relations the government was either subject to or was engaged in, including:

Border closures/disputes
Allegations of war crimes against the government
Breaking of, or re-establishing of, diplomatic relations
International embargoes against the government
Whether the government receives foreign aid
Whether any government assets were frozen
Whether the government has been accused of manipulating humanitarian aid
Whether the government is engaged in international disputes or territorial
    disputes
Whether there are travel bans against government officials

- *Government Legitimacy*—these variables relate to the legitimacy of the governments involved, including whether the governments were autocratic, had coup d'etats, or came to power through held legitimate elections.
- *Ecological aspects*—these variables relate to the ecological environment in which LeT functioned such as the state of deforestation, availability of drinking water, and occurrence of natural disasters.
- *Physical Infrastructure*—these variables describe the physical infrastructure of the countries where LeT operates, such as the presence of airports, oil and gas facilities and pipelines, border controls, ports and harbors, railways, roadways, and waterways.
- *Physical Resources*—these variables describe the availability of natural and man-made resources in the region. The natural resource variables focused on availability of various minerals and ores, while the man-made resources focused largely on arms and ammunition, automobiles and transportation vehicles, electricity/power, and medicines/pharmaceuticals.
- *Terrain*—terrain can often aid insurgents, the dataset includes variables to describe the kinds of terrain in which LeT was operating.
- *Demographics*—these variables relate to the demographics of the population, including standard demographic indicators such as fertility rates, life expectancy, male-female ratios, median age, and total population.
- *Economic Variables*—a large set of variables describes to the sources of revenue for the countries involved: agriculture, fisheries, and industry. These variables are related to food prices, GDP and per capita income, balance of revenues and expenditures, the inflation rate, and the poverty rate.

- *External Actor*—these variables describe the behaviors of external actors in the region including the presence of:

  Foreign militaries
  Foreign non-state armed groups
  Government organizations
  International businesses
  Foreign international organizations
  Peacekeeping forces

- *Security Related*—these variables cover the security situation in the region responding to issues such as:

  Is there inter-group tension?
  Are there conflicts with casualties on the side of foreign security forces in the region?
  Are there conflicts with a national security force with casualties on the security force side?
  Are there inter-clan/tribe tensions?
  Are there non-state armed groups in the region?

- *Environment Social Structure*—these variables study the social structure of the population of the region, including distributions of clans/tribes, distributions of ethnicities and forms of belief, the existence of internally displaced persons, whether the population has access to the Internet, the presence or absence of media organizations, literacy rates, and the refugee populations in the area.

- *LeT environmental variables*—Group Structure and Organization. We defined a large set of variables to study LeT itself.

  "Basic Group" variables dealt with whether the group had multiple names, the size of the organization, whether the group dissolved or split during a time frame, and information on umbrella organizations with which the group is affiliated.

  "Group Character" variables deal with whether the group is militant, political, or religious.

  "Group Equipment" variables deal with the types of armaments, vehicles, chemicals, and other equipment the group has in its possession.

  "Group Infrastructure" variables deal with how the group functions. This is a very important set of variables and includes whether the group is engaged in

  Businesses
  Charities
  Commodities
  Property businesses

In addition, this class of variables includes information on whether the group has a network structure, and whether it maintains offices and training camps.

Intra-Organizational Conflict—this set of variables examines whether the group was engaged in internal conflict and the cause of such conflict (e.g., for ideological reasons, due to differences in goals of factions, due to differences about leadership, or conflicts about resources).

Group Leadership—this class of variables relates to the type of leadership of LeT and whether those leaders were arrested, or released from arrest, whether they are charismatic, or corrupt? There are also variables related to whether the leadership is paid, whether they have military experience, whether they are spiritual, leaders and so forth.

Legitimacy of Group Leadership—these variables examine the sources of the LeT leadership's legitimacy.

Group's Local Organization—this set of variables deals with how the group operates geographically, such as whether or not they are a cross border organization, operate in displaced persons' camps or not (both internally and transnationally), or whether or not the organization's areas of operation have expanded or contracted.

Group Membership—this class of variables studies the membership of the group, and the types of populations from which the group recruits its members. It defines variables related to members' belief system, whether child soldiers and forceful recruitments are used, whether foreigners are join the group, whether the LeT uses gender as a basis to decide who to recruit, whether members of LeT are also members of other non-state armed groups, and the socio-economic status of recruits.

Group Organization Split—this set of variables examines whether the group has split, why, and whether it has reunited.

Group Support—this important set of variables checks whether LeT provided financial, material, military, or political support to another non-state armed group.

Group Aspirations & Objectives—this set of variables seeks to describe the goals and aspirations of the group.

Group Alliances—this set of variables looks at the types of relationships the group has with a wide variety of other actors including:

> Foreign governments
> International businesses
> NGOs
> Other non-state armed groups
> Security forces
> Prominent leaders

- *Group Domestic Relations*—these variables examine the group's domestic relations with the local (or national) government: did LeT members receive amnesty, were its members subject to arrest (or released from arrest), were there arrest warrants, government bans, frozen assets associated with the group? Did the government mistreat group members or kill their members arbitrarily? Did the government carry out raids and other forms of repression? What kinds of media statements did the government put out?

- *Group International Relations*—these variables describe LeT's international relationships. This includes international allegations of human rights abuses/ war crimes by LeT, arrests/extraditions of LeT members by foreign states, international bans, international designation of LeT as a terror organization, international embargos, international arrest warrants, assets freezes by international parties, international sanctions and resolutions against LeT, and travel bans on LeT personnel. A further set of variables describes ongoing negotiations involving LeT.
- *Sources of Support*—a relatively important set of variables describe support to LeT from different entities. Support variables are defined both by the type of support (financial, military, political, and material) as well as by the entity providing the support (e.g., the local population, diaspora, local government, foreign state, intergovernmental organization, non-governmental organization, non-state armed group, etc.).

A.2 Learning Behavioral Models

This section briefly describes what a *behavioral model* of LeT looks like, and then explains how these behavioral models are automatically extracted from our LeT data set.

A *behavioral model* of a terror group consists of a set of probabilistic rules of the form "When condition C is true (in the environment in which LeT is operating), then there is a probability of P % that LeT will carry out a given action A with intensity level I".

For example, we discovered about LeT the following SOMA rule (provided here solely as an example—detailed discussion is provided in previous chapters):When we examine this SOMA rule, we note that we can write it as the probabilistic logic rule given below.

> When 0–4 leaders of LeT died during a given month and LeT was receiving financial support from its diaspora, there is an 87.5 % probability that it will carry out attacks on local security forces. (Support = 14, Inverse Probability = 1, Negative Probability = 0.)

$$attackLSF(1,1) : 0.875 \leftarrow leaddeceased(0,4) \& diaspora \sup port(1,1).$$

All SOMA rules have a *head* and a *body*. We illustrate these concepts via the example given below:

- *attackLSF*(1,1):0.875 is the *head* of this rule. The head of a rule always consists of:
    - an action variable ("*attackLSF*" in this case),
    - a range of values for the *attackLSF* variable—in this case, the range of values is [1,1] indicating that this variable lies between 1 and 1 (i.e., equals 1), and
    - A probability, which in this case is 0.875.

- *leaddeceased*(0,4) & *diasporasupport*(1,1) is the *body* of this rule. In this example, the body consists of two "environmental atoms" *leaddeceased*(0.4) and *diasporasupport*(1,1) connected together by an "&" symbol that denotes logical conjunction (or "and"). In general, only environmental variables can be referenced in the body of a rule—the variables *leaddeceased* and *diasporasupport* both represent aspects of the environment in which LeT is functioning. In this case, *leaddeceased*(0,4) is true w.r.t. a particular month in the LeT data set if the leaddeceased attribute for that month has a value between 0 and 4. In contrast to *leaddeceased* which is a numeric variable, *diasporasupport* is a binary variable (we only code this variable as occurring or not).

Informally speaking, this rule says that the probability of the *attacksLSF* variable having a value of 1, given that the *leaddeceased* variable has a value in the 0–4 interval and given that the variable *diasporasupport* has value 1, is 87.5 %.

Thus, all SOMA-rules have the form

$$HeadVar(Val1, Val2) : Prob \leftarrow BodyAtom_1 \& \ldots \& BodyAtom_k$$

A general SOMA-rule may be read as: When $BodyAtom_1$ and $BodyAtom_2$ and $BodyAtom_k$ are all true in a given month, then there is a probability of *Prob* that *HeadVar* will have a value between *Val1* and *Val2* during that month.

SOMA-rules are extracted automatically using algorithms we have developed (Ernst and Subrahmanian 2004; Khuller et al. 2007). The question the reader will have is the following: given a data set such as our LeT data set, how do we decide what SOMA rules are worth extracting? For this, our SOMA rule extraction engine uses four parameters for any possible rule of the form

$$HeadVar(Val1, Val2) : Prob \leftarrow BodyAtom_1 \& \ldots \& BodyAtom_k$$

The *Support* of this rule refers to the number of months when both the head and the body of this rule were true, i.e., when the atom in the rule head and all environmental atoms in the rule body were true. The SOMA-extraction engine requires that all rules have at least a minimal, user specified support.

The *Confidence* (or *Probability*) of a rule refers to the ratio of the number of months when both the body and the head of the rule were true to the number of times just the body of the rule was true, i.e.,

$$Confidence = \frac{number\ of\ months\ when\ HeadVar(Val1, Val2)\ \&\ BodyAtom_1\ \&\ BodyAtom_2\ \&\ \ldots\ \&\ BodyAtom_k\ is\ true}{number\ of\ months\ when\ BodyAtom_1\ \&\ BodyAtom_2\ \&\ \ldots\ \&\ BodyAtom_k\ is\ true}$$

Thus, *confidence* of a rule is merely the conditional probability of LeT taking an action at a certain intensity level, given that the body of the rule is true. The SOMA rule extraction engine requires that all extracted rule pass a desired confidence threshold.

The *Inverse Probability* of a rule refers to the ratio of the number of months the head and the nody of the rule are both true to the number of months the head is true. Thus, this inverse probability is the conditional probability of the body being true, given that the head is true.

$$Inverse\ Prob = \frac{number\ of\ months\ when\ HeadVar(Val1, Val2)\ \&\ BodyAtom_1\ \&\ BodyAtom_2\ \&\ \ldots\ \&\ BodyAtom_k\ is\ true}{number\ of\ months\ when\ HeadVar(Val1, Val2)\ is\ true}$$

The SOMA rule extraction engine requires that the inverse probability exceeds a threshold. Intuitively, when both the *Probability* and the *Inverse Probability* are high, this means that the rule head and the rule body co-occur with a high degree of likelihood, thus implying a strong correlation between the two.

Finally, the *Negative Probability* of a SOMA-rule is the ratio of the number of months the head and the body of a SOMA rule are both true to the number of months the body is not true. In other words, the negative probability of a SOMA rule is merely the conditional probability of the head being true, given that the rule body is false.

$$Negative\ Prob = \frac{number\ of\ months\ when\ HeadVar(Val1, Val2)\ \&\ \sim BodyAtom_1\ \&\ BodyAtom_2\ \&\ \ldots\ \&\ BodyAtom_k\ is\ true}{number\ of\ months\ when\ \sim BodyAtom_1\ \&\ BodyAtom_2\ \&\ \ldots\ \&\ BodyAtom_k\ is\ true}$$

The SOMA rule extraction engine only extracts rules that have a negative probability below a given threshold. This is because finding conditions that have a high probability of predicting an action (when they are true) and a low probability of predicting an action (when they are false) is the ultimate goal of these forecasting engines.

Appendix B
List of All Terrorist Attacks Carried out by LeT

The list of attacks given below only includes attacks attributed to LeT by the US National Counter-Terrorism Center (NCTC) in their Worldwide Incident Tracking System (WITS). WITS does not always ascribe responsibility for an attack—in some cases, a likely perpetrator is mentioned. Those cases where LeT has been mentioned in this way are also listed below.

V. S. Subrahmanian et al., *Computational Analysis of Terrorist Groups: Lashkar-e-Taiba*,   189
DOI: 10.1007/978-1-4614-4769-6, © Springer Science+Business Media New York 2013

Date	Country	City	Details	Dead	Wounded	Hostages	Total casualties
20-Jul-04	India	Rajaur	4 civilians, 1 child killed in armed attack by suspected LT in Rajaur, Jammu and Kashmir, India	5	0	0	5
20-Jul-04	India	Doda	4 police officers killed, 1 other wounded in armed attack by suspected LT in Doda, Jammu and Kashmir, India	4	1	0	5
18-Aug-04	India	Malachamlan	1 civilian, 3 children killed in armed attack by suspected LT in Malachamlan, Jammu and Kashmir, India	4	0	0	4
12-Sep-04	India	Punch	3 civilians killed in armed attack by suspected LT in Punch, Jammu and Kashmir, India	3	0	0	3
18-Oct-04	India	Khog	1 civilian killed, 14 others wounded in grenade attack by LT in Khog, Jammu and Kashmir, India	1	14	0	15
27-Nov-04	India	Doda	2 police officers injured in armed attack by LT in Shudaan Village, Doda, Jammu and Kashmir, India	0	2	0	2
11-Jan-05	India	Partabpora	1 civilian killed in armed attack by suspected LT in Partabpora Village, Jammu and Kashmir, India	1	0	0	1
15-Mar-05	Pakistan	Faisalabad	Police officers targeted in armed attack by suspected LT in Faisalabad, Punjab, Pakistan	0	0	0	0
29-Mar-05	India	Udhampur	1 civilian killed by suspected LT in Udhampur, Jammu and Kashmir, India	1	0	0	1
17-May-05	India	Srinagar	4 civilians killed, 2 others kidnapped by suspected LT near Srinagar, Jammu and Kashmir, India	4	0	2	6
19-Jun-05	India	Mendhar	1 civilian, 2 police officers killed, 1 police officer wounded in armed attack by suspected LT in Mendhar, Jammu and Kashmir, India	3	1	0	4
5-Jul-05	India	Faizabad	2 civilians, 4 paramilitaries wounded in attack by suspected LT in Faizabad, Uttar Pradesh, India	0	6	0	6
18-Jul-05	India	Udhampur	1 government official, 5 civilians killed in armed attack by suspected LT in Udhampur, Jammu and Kashmir, India	6	0	0	6

(continued)

(continued)

Date	Country	City	Details	Dead	Wounded	Hostages	Total casualties
28-Jul-05	India	Udhampur	1 child, 2 civilians killed, 2 other civilians wounded in armed attack by suspected LT in Udhampur, Jammu and Kashmir, India	3	2	0	5
30-Aug-05	India	Baramula	1 security guard wounded in grenade attack by LT in Baramula, Jammu and Kashmir, India	0	1	0	1
2-Sep-05	India	Doda	1 government employee, 2 civilians killed, 4 civilians wounded in armed attack by suspected LT in Doda, Jammu and Kashmir, India	3	4	0	7
9-Sep-05	India	Udhampur	1 police officer wounded, 3 residences damaged in armed attack in Udhampur, Jammu and Kashmir, India	0	1	0	1
5-Oct-05	India	Udhampur	3 civilians killed in armed attack by suspected LT in Udhampur, Jammu and Kashmir, India	3	0	0	3
18-Oct-05	India	Srinagar	1 police officer killed in armed attack in Srinagar, Jammu and Kashmir, India	1	0	0	1
18-Oct-05	India	Srinagar	1 government official, 1 government employee, 1 guard killed, 1 civilian, 1 guard wounded in armed attack in Srinagar, Jammu and Kashmir, India	3	2	0	5
29-Oct-05	India	New Delhi	62 civilians killed, 210 others wounded in bombings in New Delhi, India	62	210	0	272
8-Nov-05	India	Rajauri	1 civilian, 1 police officer killed in armed attack by suspected LT in Rajauri, Jammu and Kashmir, India	2	0	0	2
16-Nov-05	India	Srinagar	4 civilians killed and 59 civilians, 12 security guards, 1 government official wounded in VBIED attack by suspected LT in Srinagar, Jammu and Kashmir, India	4	72	0	76
22-Nov-05	India	Rajauri	1 civilian kidnapped by suspected LT in Rajauri, Jammu and Kashmir, India	0	0	1	1
28-Dec-05	India	Bangalore	1 educator killed, 3 others wounded in armed attack by suspected LT in Bangalore, Karnataka, India	1	3	0	4
24-Jan-06	Pakistan	Gurjanwala	1 civilian kidnapped by suspected LT in Gurjanwala, Punjab, Pakistan	0	0	1	1

(continued)

Date	Country	City	Details	Dead	Wounded	Hostages	Total casualties
(continued)							
7-Mar-06	India	Varanasi	15 civilians killed, 101 others wounded in bombing in Varanasi, Uttar Pradesh, India	15	101	0	116
29-Mar-06	India	Anantnag	1 politically affiliated entertainer kidnapped and killed in armed attack by LT in Anantnag, Jammu and Kashmir, India	1	0	0	1
6-Apr-06	India	Baramula	1 vehicle damaged in armed attack by LT in Baramula, Jammu and Kashmir, India	0	0	0	0
20-Apr-06	India	Doda	1 police officer killed, 3 others injured in grenade attack by suspected LT in Doda, Jammu and Kashmir, India	1	3	0	4
22-Apr-06	India	Pulwama	Soldiers targeted in VBIED explosion in Pulwama, Jammu and Kashmir, India	0	0	0	0
1-May-06	India	Doda	22 civilians killed, 6 others injured in kidnapping and armed attack by LT in Doda, Jammu and Kashmir, India	22	6	0	28
8-May-06	India	Pulwama	1 civilian kidnapped and killed in armed attack by suspected LT in Pulwama, Jammu and Kashmir, India	1	0	0	1
13-May-06	India	Doda	2 political activists killed, 34 others, 1 child injured in grenade attack in Doda, Jammu and Kashmir, India	2	35	0	37
21-May-06	India	Srinagar	3 political workers, 2 police officers killed, 14 civilians, 9 police officers, 2 political workers wounded in armed attack in Srinagar, Jammu and Kashmir, India	5	25	0	30
22-May-06	India	Srinagar	Police officer targeted in blood poisoning attack in Srinagar, Jammu and Kashmir, India	0	0	0	0
24-May-06	India	Srinagar	1 hospital damaged in arson in Srinagar, Jammu and Kashmir, India	0	0	0	0
1-Jun-06	India	Nagpur	2 police officers wounded in armed attack by suspected LT in Nagpur, Maharashtra, India	0	2	0	2
12-Jun-06	India	Kulgam	8 civilians, 1 child, 1 soldier killed, 4 civilians wounded in kidnapping, assault, and armed attack in Kulgam, Jammu and Kashmir, India	10	4	0	14

(continued)

(continued)

Date	Country	City	Details	Dead	Wounded	Hostages	Total casualties
12-Jun-06	India	Srinagar	1 civilian killed, 29 others wounded in grenade attack by suspected LT in Srinagar, Jammu and Kashmir, India	1	29	0	30
22-Jun-06	India	Baramula	3 civilians killed, 14 others, 1 religious healer wounded in grenade attack by suspected LT in Baramula, Jammu and Kashmir, India	3	15	0	18
1-Jul-06	India	Udhampur	3 civilians kidnapped, 6 paramilitaries held hostage by suspected LT in Udhampur, Jammu and Kashmir, India	0	0	9	9
4-Jul-06	India	Punch	1 civilian kidnapped and killed in assault by LT in Punch, Jammu and Kashmir, India	1	0	0	1
11-Jul-06	India	Srinagar	2 civilians killed, 6 others wounded in grenade attack by suspected LT in Srinagar, Jammu and Kashmir, India	2	6	0	8
11-Jul-06	India	Mumbai	203 civilians, 6 children killed, 863 civilians, 27 children wounded in IED attacks in Mumbai, Maharashtra, India	209	890	0	1099
11-Jul-06	India	Srinagar	7 civilians wounded in grenade attack by suspected LT in Srinagar, Jammu and Kashmir, India	0	7	0	7
11-Jul-06	India	Srinagar	5 civilians, 2 children, 1 soldier wounded in grenade attack by suspected LT in Srinagar, Jammu and Kashmir, India	0	8	0	8
11-Jul-06	India	Srinagar	15 civilians wounded in grenade attack by LT in Srinagar, Jammu and Kashmir, India	0	15	0	15
11-Jul-06	India	Srinagar	6 civilians killed, 12 others wounded in grenade attack by suspected LT in Srinagar, Jammu and Kashmir, India	6	12	0	18
12-Jul-06	India	Doda	1 paramilitary kidnapped and killed by suspected LT in Doda, Jammu and Kashmir, India	1	0	0	1
12-Jul-06	India	Srinagar	8 civilians wounded in grenade attack by suspected LT in Srinagar, Jammu and Kashmir, India	0	8	0	8
15-Jul-06	India	Srinagar	1 civilian killed, 10 others wounded in grenade attack in Srinagar, Jammu and Kashmir, India	1	10	0	11

(continued)

(continued)

Date	Country	City	Details	Dead	Wounded	Hostages	Total casualties
17-Jul-06	India	Punch	1 police officer killed in armed attack by suspected LT in Punch, Jammu and Kashmir, India	1	0	0	1
19-Jul-06	India	Punch	1 police officer killed in armed attack by LT in Punch, Jammu and Kashmir, India	0	0	0	0
23-Jul-06	India	Punch	1 police station damaged in armed attack at Punch, Jammu and Kashmir, India	0	0	0	0
25-Jul-06	India	Doda	1 child killed in assault by suspected LT in Doda, Jammu and Kashmir, India	1	0	0	1
10-Aug-06	India	Udhampur	1 civilian, 2 children killed, 1 civilian kidnapped in armed attack by LT in Udhampur, Jammu and Kashmir, India	3	0	1	4
8-Sep-06	India	Nasik	31 civilians killed, 125 wounded in IED attacks in Nasik, Maharashtra, India	31	125	0	156
8-Sep-06	India	Srinagar	1 civilian killed, 1 child wounded in armed attack by suspected LT in Srinagar, Jammu and Kashmir, India	1	1	0	2
7-Oct-06	India	Baramula	1 dentist kidnapped and killed in assault by suspected LT in Baramula, Jammu and Kashmir, India	1	0	0	1
8-Oct-06	India	Baramula	1 vehicle destroyed in arson by suspected LT in Baramula, Jammu and Kashmir, India	0	0	0	0
28-Oct-06	India	Udhampur	3 children, 1 civilian kidnapped by LT in Udhampur, Jammu and Kashmir, India	0	0	4	4
21-Dec-06	India	Baramula	1 civilian kidnapped and killed in armed attack by LT in Baramula, Jammu and Kashmir, India	1	0	0	1
30-Dec-06	India	Pulwama	1 police officer, 1 civilian killed, 1 police officer, 2 civilians wounded in armed attack by LT in Pulwama, Jammu and Kashmir, India	2	3	0	5
7-Jan-07	India	Baramula	1 political affiliate killed in armed attack by LT in Baramula, Jammu and Kashmir, India	1	0	0	1

(continued)

(continued)

Date	Country	City	Details	Dead	Wounded	Hostages	Total casualties
9-Feb-07	India	Pulwama	1 civilian killed in armed attack by LT in Pulwama, Jammu and Kashmir, India	1	0	0	1
18-Feb-07	India	Panipat	57 civilians, 8 children, 3 police officers killed, 46 civilians, 4 children wounded in IED attacks in Panipat, Haryana, India	68	50	0	118
30-Mar-07	India	Rajauri	5 civilians kidnapped and killed, 4 others kidnapped and wounded, 2 others kidnapped in armed attack by LT in Rajauri, Jammu and Kashmir, India	5	4	2	11
20-Apr-07	India	Doda	1 civilian killed in armed attack by LT in Doda, Jammu and Kashmir, India	1	0	0	1
21-Apr-07	India	Doda	3 civilians killed in assault by LT in Doda, Jammu and Kashmir, India	3	0	0	3
18-May-07	India	Hyderabad	12 civilians killed, 50 others wounded in IED attack in Hyderabad, Andhra Pradesh, India	12	50	0	62
1-Jun-07	India	Baramula	1 police officer, 1 soldier killed, 10 soldiers, 6 police officers wounded in armed attack by LT in Baramula, Jammu and Kashmir, India	2	16	0	18
9-Jun-07	India	Pulwama	2 paramilitaries, 2 civilians wounded in armed attack by LT in Pulwama, Jammu and Kashmir, India	0	4	0	4
15-Jun-07	India	Baramula	1 soldier, 1 paramilitary member, 1 civilian killed, 5 civilians, 4 soldiers wounded in armed attack by LT in Baramula, Jammu and Kashmir, India	3	9	0	12
25-Jun-07	India	Doda	2 civilians killed, 14 others wounded in grenade attack by suspected LT in Doda, Jammu and Kashmir, India	2	14	0	16
2-Jul-07	India	Doda	1 police officer killed, 5 wounded in armed attack by suspected LT in Doda, Jammu and Kashmir, India	1	5	0	6
18-Jul-07	India	Doda	1 civilian killed in grenade attack by suspected LT in Doda, Jammu and Kashmir, India	1	0	0	1
30-Jul-07	India	Doda	2 civilians kidnapped by suspected LT in Doda, Jammu and Kashmir, India	0	0	2	2

(continued)

(continued)

Date	Country	City	Details	Dead	Wounded	Hostages	Total casualties
10-Sep-07	India	Doda	2 civilians wounded in assault by suspected LT in Doda, Jammu and Kashmir, India	0	2	0	2
17-Sep-07	India	Doda	6 children kidnapped by suspected LT in Doda, Jammu and Kashmir, India	0	0	6	6
21-Jan-08	India	Kupwara	1 police officer killed in assault by suspected LT in Kupwara, Jammu and Kashmir, India	1	0	0	1
23-Jan-08	India	Doda	1 civilian kidnapped and killed in armed attack by suspected LT in Doda, Jammu and Kashmir, India	1	0	0	1
13-Jun-08	India	Kishtwar	3 soldiers, 2 government employees killed in armed attack by LT in Kishtwar, Jammu and Kashmir, India	5	0	0	5
25-Jul-08	India	Bangalore	2 civilians killed, 20 wounded in IED attack in Bangalore, Karnataka, India	2	20	0	22
16-Sep-08	India	Doda	1 civilian kidnapped and wounded, 11 civilians wounded in assault by suspected LT in Doda, Jammu and Kashmir, India	0	11	1	12
20-Nov-08	India	Baramula	1 police officer wounded in RPG attack by LT in Baramula, Jammu and Kashmir, India	0	1	0	1
26-Nov-08	India	Mumbai	145 civilians, 17 police officers, 2 military members killed; 277 civilians, 3 government employees, 3 police officers, 25 military members wounded in assault, bombings, IED and armed attacks in Mumbai, Maharashtra, India	164	308	0	472
10-Apr-09	India	Rajauri	1 police officer kidnapped and killed in assault by suspected LT in Rajauri, Jammu and Kashmir, India	1	0	0	1
18-Apr-09	India	Doda	1 civilian, 1 child kidnapped by suspected LT in Doda, Jammu and Kashmir, India	0	0	2	2
21-Apr-09	India	Punch	5 civilians, 1 child killed, 7 civilians wounded in IED attack by LT in Punch, Jammu and Kashmir, India	6	7	0	13

(continued)

(continued)

Date	Country	City	Details	Dead	Wounded	Hostages	Total casualties
12-Sep-09	India	Srinagar	3 police officers, 1 civilian killed, 9 police officers wounded in VBIED attack by suspected LT in Srinagar, Jammu and Kashmir, India	4	9	0	13
28-Sep-09	India	Pulwama	1 civilian killed, 4 paramilitary members wounded in armed attack by suspected LT in Pulwama, Jammu and Kashmir, India	1	4	0	5
6-Jan-10	India	Srinagar	3 civilians, 1 police officer, 1 paramilitary member killed, 9 civilians, 3 paramilitary members wounded in armed attack in Srinagar, Jammu and Kashmir, India	5	12	0	17
13-Feb-10	India	Pune	17 civilians killed, 65 others wounded in IED attack in Pune, Maharashtra, India	17	60	0	77
26-Feb-10	Afghanistan	Kabul	14 civilians, 3 police officers, 2 soldiers, 1 diplomat killed; 32 civilians, 6 police officers, 5 soldiers wounded in suicide IED, VBIED, and armed attacks in Kabul, Afghanistan	20	43	0	63
25-Apr-10	India	Amritsar	2 police officers killed in armed attack by suspected LT in Amritsar, Punjab, India	2	0	0	2
10-Aug-10	India	Baramula	3 police officers killed in armed attack by suspected LT in Baramula, Jammu and Kashmir, India	3	0	0	3
11-Aug-10	India	Rajauri	2 civilians killed, 18 others, 2 paramilitary officers wounded in armed attack by suspected LT in Rajauri, Jammu and Kashmir, India	2	20	0	22
21-Nov-10	India	Kishtwar	1 civilian killed, 1 civilian, 2 children wounded in armed attack by suspected LT in Kishtwar, Jammu and Kashmir, India	1	3	0	4
31-Jan-11	India	Baramula	1 child, 1 civilian kidnapped and killed in armed attack, 1 child wounded in assault by suspected LT in Baramula, Jammu and Kashmir, India	2	1	0	3
16-May-11	India	Kishtwar	1 political party leader kidnapped and killed in assault by suspected LT in Kishtwar, Jammu and Kashmir, India	1	0	0	1
27-May-11	India	Kupwara	2 civilians killed in armed attack by suspected LT in Kupwara, Jammu and Kashmir, India	2	0	0	2

(continued)

(continued)

Date	Country	City	Details	Dead	Wounded	Hostages	Total casualties
24-Jun-11	India	Srinagar	5 civilians, 1 police officer wounded in grenade attack by suspected LT in Srinagar, Jammu and Kashmir, India	0	6	0	6
30-Jun-11	India	Kishtwar	1 police officer wounded in grenade attack by suspected LT in Kishtwar, Jammu and Kashmir, India	0	1	0	1

Appendix C
List of all Temporal-Probabilistic Rules Presented in this Book

V. S. Subrahmanian et al., *Computational Analysis of Terrorist Groups: Lashkar-e-Taiba*,   199
DOI: 10.1007/978-1-4614-4769-6, © Springer Science+Business Media New York 2013

Rule name	Time offset	Dependent variable	Lower bound	Upper bound	Support	Probability	Inverse probability	Negative probability	Independent variable #1	Lower bound	Upper bound	Independent variable #2	Lower bound	Upper bound
AAH-1	1	LeT attacks civilians	1	1	10	100	100	100	LeT religious	1	1	LeT commanders died	0	2
AAH-2	3	LeT attacks Hindus	1	9	10	90.9	90.9	33.3	LeT made territorial claims	0	0	LeT commanders died	0	2
AAH-3	1	LeT attacks civilians	1	3	10	100	100	0	LeT religious	1	1	Raids on LeT	0	12
AAH-4	1	LeT attacks civilians	1	3	10	100	100	0	LeT religious	1	1	LeT personnel arrested	0	6
AAH-5	2	LeT attacks Hindus	1	9	10	90.9	90.9	33.3	LeT made territorial claims	1	1	LeT personnel arrested	0	4
AAH-6	2	LeT attacks civilians	1	3	11	100	100	0	LeT basic organization dissolved	0	0	Arrest warrants for LeT issued	1	39
AAH-7	3	LeT attacks Hindus	1	9	10	90.9	90.9	33.3	LeT made territorial claims	0	0	Arrest warrants for LeT issued	0	7
AAH-8	1	LeT attacks civilians	1	3	10	100	100	0	LeT religious	1	1	LeT personnel killed by government	0	10
AAH-9	1	LeT attacks civilians	1	3	10	100	100	0	LeT religious	1	1	Partial government ban	1	1
AAH-10	1	LeT attacks civilians	1	3	10	100	100	0	LeT religious	1	1	LeT splintering	0	0
AAH-11	1	LeT attacks Hindus	1	3	10	100	100	0	LeT religious	1	1	LeT involved in intra-org conflict	0	0
AAH-12	2	LeT attacks civilians	1	3	10	100	100	0	LeT splintering	0	0	LeT personnel arrested	1	1
AAH-13	3	LeT attacks Hindus	1	1	21	95.5	100	0	LeT splintering	0	0		0	0
AAH-14	1	LeT attacks Hindus	1	1	15	100	100	0	LeT commanders died	0	0	LeT members deserted	1	3
PST-1	3	LeT attacks symbolic sites	0	29	100	100	100	0	LeT splintering	0	0		0	0

(continued)

(continued)

Rule name	Time offset	Dependent variable	Lower bound	Upper bound	Support	Probability	Inverse probability	Negative probability	Independent variable #1	Lower bound	Upper bound	Independent variable #2	Lower bound	Upper bound
PST-2	3	LeT attacks symbolic sites	0	0	40	90.9	100	0	LeT locations across the border in India	1	1	LeT commanders died	0	5
PST-3	2	LeT attacks symbolic sites	0	0	40	97.6	95.2	40	LeT locations across the border in India	1	1	LeT commanders died	0	4
PST-4	3	LeT attacks symbolic sites	0	0	40	90.9	100	0	LeT locations across the border in India	1	1	LeT personnel released by government	0	9
PST-5	3	LeT attacks symbolic sites	0	0	40	90.9	100	0	LeT locations across the border in India	1	1	LeT personnel killed by government	0	15
PST-6	3	LeT attacks transportation facilities	0	0	106	86.9	100	0	LeT commanders died	0	6			
PST-7	3	LeT attacks transportation facilities	0	0	104	87.4	98.1	0	LeT commanders died	0	6	LeT personnel killed by government	0	14
PSF-1	2	LeT attacks professional security forces	1	1	14	87.5	100	0	LeT received military support from Pakistani government	1	1			
PSF-2	2	LeT attacks professional security forces	1	1	13	92.9	92.9	33.3	Government ban on LeT	0	0	LeT has strong relationship with Pakistani military	1	1
PSF-3	3	LeT attacks professional security forces	1	1	14	87.5	93.3	33.3	LeT waging communications campaign using periodicals	1	1			
PSF-4	2	LeT attacks professional security forces	1	1	17	89.5	68	47.1	LeT waging communications campaign using periodicals	1	1	LeT personnel arrested	0	2

(continued)

(continued)

Rule name	Time offset	Dependent variable	Lower bound	Upper bound	Support	Probability	Inverse probability	Negative probability	Independent variable #1	Lower bound	Upper bound	Independent variable #2	Lower bound	Upper bound
PSF-5	2	LeT attacks professional security forces	1	1	23	88.5	92	20	LeT waging communications campaign using periodicals	1	1	LeT commanders died	0	0
PSF-6	3	LeT attacks professional security forces	1	1	14	93.3	93.3	25	LeT waging communications campaign using periodicals	1	1	LeT personnel killed by government	0	5
PSF-7	3	LeT attacks professional security forces	1	1	15	88.2	100	0	LeT waging communications campaign using periodicals	1	1	LeT personnel killed by government	0	5
PSF-8	3	LeT attacks professional security forces	1	1	12	92.3	75	44.4	LeT's leadership functionally differentiated	1	1	LeT personnel released by government	1	2
PSF-9	3	LeT attacks professional security forces	0	0	11	91.7	91.7	25	LeT members deserted	1	1			
PSF-10	1	LeT attacks local security installations	0	0	99	90	100	0	Arrest warrants for LeT issued	1	39	LeT splintering	0	0
PSF-11	1	LeT attacks professional security forces	0	1	29	87.9	100	0	LeT personnel killed by government	0	16	LeT splintering	0	0
SI-1	2	LeT attacks Indian security installations	1	1	12	92.3	100	0	LeT waging communications campaign using periodicals	1	1	LeT personnel arrested	0	0
SI-2	2	LeT attacks Indian security installations	1	1	10	100	83.3	28.6	LeT waging communications campaign using periodicals	1	1	LeT splintering	0	0
SI-3	3	LeT attacks Indian security installations	1	1	14	87.5	77.8	21.1	LeT waging communications campaign using periodicals	1	1	LeT personnel arrested	0	0

(continued)

(continued)

Rule name	Time offset	Dependent variable	Lower bound	Upper bound	Support	Probability	Inverse probability	Negative probability	Independent variable #1	Lower bound	Upper bound	Independent variable #2	Lower bound	Upper bound
SI-4	2	LeT attacks Indian security installations	1	1	12	92.3	100	0	LeT waging communications campaign using periodicals	1	1	Government ban on LeT	0	0
SI-5	2	LeT attacks Indian security installations	1	1	12	92.3	100	0	LeT waging communications campaign using periodicals	1	1	LeT personnel arrested	0	19
SI-6	2	LeT attacks Indian security installations	1	1	11	100	91.7	16.7	LeT waging communications campaign using periodicals	1	1	LeT assets frozen by the government	0	0
SI-7	2	LeT attacks Indian security installations	1	3	10	90.9	100	0	LeT provides support to other Islamist organizations	1	1	LeT engaged in intra-organizational conflict	0	0
SI-8	2	LeT attacks Indian security installations	1	3	10	90.9	100	0	LeT received military support from Pakistani government	1	1	LeT leaders resigned	0	0
AOH-1	2	LeT attacks on a holiday	1	1	11	91.7	100	0	LeT sought to remove Indian influence from Kashmir	1	1	LeT personnel arrested	0	0
AOH-2	2	LeT attacks on a holiday	1	1	11	91.7	100	0	LeT sought to remove Indian influence from Kashmir	1	1	LeT personnel killed by government	0	8
AOH-3	2	LeT attacks on a holiday	1	1	10	90.9	90.9	33.3	LeT personnel arrested	0	8			
AOH-4	2	LeT attacks on a holiday	1	1	11	91.7	100	0	LeT advocates a change in lifestyle	1	1	LeT personnel arrested	0	0
AOH-5	2	LeT attacks on a holiday	1	1	11	91.7	100	0	LeT advocates a change in lifestyle	1	1	LeT personnel killed by government	0	8

(continued)

(continued)

Rule name	Time offset	Dependent variable	Lower bound	Upper bound	Support	Probability	Inverse probability	Negative probability	Independent variable #1	Lower bound	Upper bound	Independent variable #2	Lower bound	Upper bound
AA-1	2	LeT attempted attacks	1	2	24	96	100	0	LeT leaders resigned	0	0			
AA-2	3	LeT attempted attacks	1	2	21	95	100	0	Arrest warrants for LeT issued	1	34			
AA-3	1	LeT attempted attacks	1	2	24	96	100	0	Raids on LeT	0	2			
AA-4	1	LeT attempted attacks	1	2	24	96	100	0	LeT personnel released by government	0	2			
AC-1	1	LeT involved with armed clashes with local security forces (with LeT casualties)	1	1	10	100	90.9	0	LeT had a campaign focused on sympathy and identity in politics	1	1	LeT personnel tried in Australia	1	1
AC-2	2	LeT involved with armed clashes with local security forces (with LeT casualties)	1	1	15	88.2	75	26.3	LeT personnel arrested	5	24	LeT personnel tried in India or Pakistan	1	1
AC-3	1	LeT involved with armed clashes with local security forces (with LeT casualties)	1	1	11	91.7	100	0	LeT members were also members of other non-state armed groups	1	1	LeT personnel killed by government	0	8
AC-4	1	LeT involved with armed clashes with local security forces (with LeT casualties)	1	1	10	100	90.9	33.3	LeT members were also members of other non-state armed groups	1	1	LeT personnel arrested	5	61

(continued)

(continued)

Rule name	Time offset	Dependent variable	Lower bound	Upper bound	Support	Probability	Inverse probability	Negative probability	Independent variable #1	Lower bound	Upper bound	Independent variable #2	Lower bound	Upper bound
AC-5	2	LeT involved with armed clashes with local security forces (with LeT casualties)	1	1	11	100	100	0	LeT is allied with Pakistan's security forces	1	1	LeT splintering	0	0
AC-6	2	LeT involved with armed clashes with local security forces (with LeT casualties)	1	1	11	91.7	84.6	20	LeT was practicing as a charitable organization	1	1	LeT personnel arrested	6	24
AC-7	3	Members of LeT are killed in Jammu and Kashmir	0	13	16	88.9	100	0	LeT provides social service medical programs	1	1	Government ban on LeT	0	0
AC-8	1	LeT involved with armed clashes with local security forces (with LeT casualties)	1	1	10	100	90.9	33.3	LeT members were also members of other non-state armed groups	1	1	LeT commanders died	0	1
AC-9	2	LeT involved with armed clashes with local security forces (with LeT casualties)	1	1	58	86.6	78.4	40	LeT commanders died	1	1			
AC-10	3	Members of LeT are killed in Jammu and Kashmir	0	13	15	93.8	100	0	LeT received military support from Pakistani government	1	1			

(continued)

(continued)

Rule name	Time offset	Dependent variable	Lower bound	Upper bound	Support	Probability	Inverse probability	Negative probability	Independent variable #1	Lower bound	Upper bound	Independent variable #2	Lower bound	Upper bound
AC-11	1	LeT involved with armed clashes with local security forces (with LeT casualties)	1	1	31	86.5	97	33.3	LeT had training camps	1	23			
AC-12	1	LeT involved with armed clashes with local security forces (with LeT casualties)	1	1	32	91.2	93.9	33.3	LeT had training camps	1	25	LeT personnel killed by government	0	11

Appendix D
List of all Policies Generated by
the Policy Computation Engine

Independent variable	P1	P2	P3	P4	P5	P6	P7	P8
LeT religious organization	–	–	–	–	–	–	–	–
LeT commanders died	X	X	X	X	X	X	X	X
LeT did not make territorial claims	–	–	–	–	–	–	–	–
Raids on LeT	X	X	X	X	X	X	X	X
LeT personnel arrested	X	X	X	X	X	X	X	X
LeT made territorial claims	–	–	–	–	–	–	–	–
LeT basic organization was not dissolved	–	–	–	–	–	–	–	–
Arrest warrants for LeT issued	X	X	X	X	X	X	X	X
LeT personnel killed by government	X	X	X	X	X	X	X	X
Partial government ban	X	X	X	X	X	X	X	X
LeT was not splintering	X	X	X	X	X	X	X	X
LeT was not involved in intra-organizational conflict	X	X	X	X	X	X	X	X
No LeT commanders died	–	–	–	–	–	–	–	–
LeT members deserted	X	X	X	X	X	X	X	X
LeT received military support from Pakistani government	X	X	X	X	X	X	X	X
No Government ban on LeT	X	X	X	X	–	–	–	–
LeT had a strong relationship with Pakistani military	–	–	–	–	X	X	X	X
LeT waging communications campaign using periodicals	X	X	X	X	X	X	X	X
LeT's leadership functionally differentiated	–	–	–	–	–	–	–	–
LeT personnel released by government	X	X	X	X	X	X	X	X
No LeT personnel arrested	–	–	–	–	–	–	–	–
No LeT assets frozen by the government	–	–	–	–	–	–	–	–
LeT provides support to other Islamist organizations	X	–	–	X	X	–	X	–
LeT was not engaged in intra-organizational conflict	–	X	X	–	–	X	–	X
No LeT leaders resigned	X	X	X	X	X	X	X	X
LeT sought to remove Indian influence from Kashmir	–	–	–	–	–	–	–	–
LeT advocates a change in lifestyle	–	–	–	–	–	–	–	–
LeT had a campaign focused on sympathy and identity in politics	X	–	X	–	X	–	–	X
LeT personnel tried in Australia	–	X	–	X	-	X	X	–
LeT personnel tried in India or Pakistan	–	–	–	–	–	–	–	–
LeT members were also members of other non-state armed groups	–	–	–	–	–	–	–	–
LeT is allied with Pakistan's security forces	–	–	–	–	–	–	–	–
LeT was practicing as a charitable organization	–	–	–	–	–	–	–	–
LeT provides social service medical programs	–	–	–	–	X	X	X	X
LeT had training camps	X	X	X	X	X	X	X	X

V. S. Subrahmanian et al., *Computational Analysis of Terrorist Groups: Lashkar-e-Taiba*, 207
DOI: 10.1007/978-1-4614-4769-6, © Springer Science+Business Media New York 2013

Appendix E
Some Background Reports
on Internal Cohesion of LeT

Reports of LeT Splits

Date	Description	Source
9/01	Link is broken	http://www.alertnet.org/thenews/newsdesk/LL367886.htm
12/01	"…renamed the Markaz Dawatul Arshad the Jamaatud Dawa and separated the Lashkar-e-Taiba infrastructure from the party. Many of his colleagues including LT chief Zakiur Rehman Lakhvi disapproved of the decision because this put the JD in control of all funds collected from abroad and locally."	http://www.dailytimes.com.pk/default.asp?page=story_18-7-2004_pg7_20
7/04	Iqbal left group founded KN	http://www.dailytimes.com.pk/default.asp?page=story_18-7-2004_pg7_20
12/09	Analyst claims attacks in Punjab was perpetrated by TTP & LeT splinter	http://www.examiner.com/south-asia-foreign-policy-in-washington-dc/lahore-attacks-and-the-splinter-ing-pakistan-s-side
2/10	STRATFOR claims that LeT has spawned splinter factions	http://news.rediff.com/report/2010/feb/25/not-much-change-in-pak-strategy-says-top-think-tank.htm
4/10	Pakistani officials claim LeT in Afghanistan are splinters from main group	http://uk.reuters.com/article/2010/04/23/us-pakistan-lashkar-idUKTRE63M2P520100423?feedType=RSS&feedName=everything&virtualBrandChannel=11708
6/10	Report that LeT hardliners have splintered from main group and gone to Wazirstan	http://www.ndtv.com/news/world/let-expands-attacks-in-afghanistan-32004.php?utm_source=twitterfeed&utm_medium=twitter&utm_campaign=Feed%3A+NdtvNews-TopStories+%28NDTV+News+-+Top+Stories%29
11/10	Report that ISI thought LeT was too powerful and split it, and LeT members wanted to fight in Afghanistan	http://www.dailytimes.com.pk/default.asp?page=2010\11\02\story_2-11-2010_pg7_13

Reports of LeT Intra-Organizational Conflicts

Date	Description	Source
12/01	"Hafiz Saeed also came under fire in December 2001, when he renamed the Markaz Dawatul Arshad the Jamaatud Dawa from and separated the Lashkar-e-Taiba infrastructure from the party. Many of his colleagues including LT chief Zakiur Rehman Lakhvi disapproved of the decision because this put the JD in control of all funds collected from abroad and locally."	http://www.dailytimes.com.pk/default.asp?page=story_18-7-2004_pg7_20
3/03	Breakaway LeT members opposed Musharraf's post 9/11 strategy,	http://news.bbc.co.uk/2/hi/south_asia/3181925.stm
7/04	Related to the Khairun Naas split	http://www.hindujagruti.org/news/5937.html
9/04	Reports on infighting, over money and caste	http://www.crisisgroup.org/~/media/Files/asia/south-asia/pakistan/095_the_state_of_sectarianism_in_pakistan.pdf
10/04	Same as 9/04	http://www.crisisgroup.org/~/media/Files/asia/south-asia/pakistan/095_the_state_of_sectarianism_in_pakistan.pdf
2/05	Reports of infighting, related to responses to Pakistani government pressure	http://www.jamestown.org/single/?no_cache=1&tx_ttnews%5Bswords%5D=8fd5893941d69d0be3f378576261ae3e&tx_ttnews%5Bany_of_the_words%5D=Lashkar&tx_ttnews%5Bpointer%5D=9&tx_ttnews%5Btt_news%5D=27599&tx_ttnews%5BbackPid%5D=7&cHash=795bc52c1527bff00b88eb2bd25b7cd5
7/09	Unclear, original link broken–this is from the same source	http://www.austlii.edu.au/cgi-bin/sinodisp/au/legis/cth/num_reg_es/ccar2009l0n126o2009414.html?stem=0&synonyms=0&query=ccar2009l0n126o2009414
9/09	LeT source told NY Times conflict between Saeed and Lakhvi over carrying out more attacks on India	http://www.nytimes.com/2009/09/30/world/asia/30mumbai.html?pagewanted=2&_r=1&ref=world
10/09	"Despite the rumours of friction between the LeT and the JuD leadership, the two segments operate in unison in South Punjab." This technically codes an absence of conflict.	http://www.newsline.com.pk/NewsSep2009/coverstorysep.htm (Link broken)

Appendix F
Some Reports of Pakistani Government Military Support for LeT

V. S. Subrahmanian et al., *Computational Analysis of Terrorist Groups: Lashkar-e-Taiba*, 213
DOI: 10.1007/978-1-4614-4769-6, © Springer Science+Business Media New York 2013

Date	Description	Source
1/01 – 7/01	*NY Times* reports that Pakistan provided support to LeT (and other militants) but stopped after 9/11	http://www.nytimes.com/2009/11/26/world/asia/26mumbai.html
12/01	Indian officials state LeT receives ISI support, Parliament attackers received ISI training	http://www.csmonitor.com/2001/1217/p7s2-wome.html
3/02	B. Raman reports that Pakistani military released LeT operatives after Musharraf ban and then helped move them to other locations in Pakistan	http://www.rediff.com/news/2002/may/17guest.htm
10/02	Specific statement by Sabahuddin Ahmed, accused military of aiding LeT training	http://news.rediff.com/special/2009/nov/10/revealed-the-dossier-on-2611-accused-sabahuddin.htm
9–10/08	Kasab stated that Pakistani Marines trained the LeT team that struck Mumbai LeT infiltration effort receives Pakistani covering fire	http://www.indianexpress.com/news/kasab-trained-by-pak-marines-report/487718/ http://satp.org/satporgtp/countries/india/states/jandk/terrorist_outfits/lashkar_e_toiba_lt.htm
7/09	ISI officers continue to provide money, weapons, and training to LeT	http://articles.timesofindia.indiatimes.com/2009-07-04/pakistan/28180045_1_al-qaida-bruce-riedel-mohammed-yahya-mujahid http://online.wsj.com/article/SB124820050290269319.html
9/09	Pakistani military continuing to help LeT infiltrate J&K	http://www.longwarjournal.org/archives/2009/09/lashkaretaiba_chief.php
10/09	Report of rogue elements of ISI & military support al-Qaeda	http://news.rediff.com/report/2009/nov/06/lashkar-is-a-jihadi-frankenstein-today.htm
2/10	Various reports ISI is working with LeT to plan attack on Mumbai	http://www.indianexpress.com/news/isi-let-getting-indian-jihadis-together-in-karachi-for-attack/573878/0 http://zeenews.india.com/news/nation/pak-training-indians-for-terror-project-report_600589.html
5/10	Authorities providing security to HSM after TTP threatens him for condemning suicide bombings	http://www.dnaindia.com/world/report_tehrik-e-taliban-pakistan-threatens-jamaat-ud-dawa-chief_1385192

Appendix G
Some Reports of LeT Using the Press
to Propagate its Message

V. S. Subrahmanian et al., *Computational Analysis of Terrorist Groups: Lashkar-e-Taiba*,   215
DOI: 10.1007/978-1-4614-4769-6, © Springer Science+Business Media New York 2013

Date	Description	Source
2/90	Report of publication of LeT magazine *Majjala-al Da'awa*	*A to Z of Jehadi Organizations in Pakistan*, Amir Rana
9/90	Report of publication of LeT magazine *Majjala-al Da'awa*	*A to Z of Jehadi Organizations in Pakstan*, Amir Rana
4/98	Saeed gives a public statement to reporters	http://www.thehindu.com/opinion/lead/article63854.ece?homepage=true
3/99	Zafar Iqbal gives a press conference	http://www.ict.org.il/NewsCommentaries/Commentaries/tabid/69/Articlsid/554/currentpage/3/Default.aspx
4/99	Zaki-ur-Rehman Lakhvi gives an interview to *The Nation*	http://www.ict.org.il/NewsCommentaries/Commentaries/tabid/69/Articlsid/554/currentpage/3/Default.aspx
8/99	LeT sent statement to AP claiming terror attack	GTD incidents: 199908150001 & 199908150002 http://www.start.umd.edu/gtd/
11/99		http://pakobserver.net/200904/08/Articles05.asp (Link broken)
4/00	Taiba bulletin published	http://www.nefafoundation.org/miscellaneous/ekeletwitnessreport.pdf
5/00	Taiba bulletin published	http://www.nefafoundation.org/miscellaneous/ekeletwitnessreport.pdf
7/00–12/00	Taiba bulletin published	http://www.nefafoundation.org/miscellaneous/ekeletwitnessreport.pdf
1/01–12/01	Various reports of LeT publications throughout 2001	*A to Z of Jehadi Organizations in Pakistan*, Amir Rana http://satp.org/satporgtp/countries/india/states/jandk/terrorist_outfits/lashkar_e_toiba.htm
7/02	Reference to LeT periodicals	*Terror in the Name of God*, Jessica Stern
5/03	Saeed gives an interview to *The Nation*	http://www.satp.org/satporgtp/countries/india/states/jandk/terrorist_outfits/lashkar_e_toiba_lt2007.htm
6/04	Reference to publication of monthly and weekly periodicals	http://www.jamestown.org/publications_details.php?volume_id=411&issue_id=3242&article_id=2369321
10/05	LeT front makes statement to Kashmiri papers	http://www.taipeitimes.com/News/front/archives/2005/10/31/2003278120
12/05	Reference to LeT publication of *Zarb-e-Momin*	http://www.satp.org/satporgtp/countries/india/states/jandk/terrorist_outfits/lashkar_e_toiba_lt2007.htm
10/06	Reference to LeT publication *Ausaf*	http://www.satp.org/satporgtp/countries/india/states/jandk/terrorist_outfits/lashkar_e_toiba_lt2007.htm
3/07	Reference to LeT publication *Bab al-Islam*	http://thecst.org.uk (link broken)

(continued)

(continued)

Date	Description	Source
5/07	Reference to multiple LeT publications, and plans to resume publication under new names in response to possible government ban	http://www.satp.org/satporgtp/countries/india/states/jandk/terrorist_outfits/lashkar_e_toiba_lt2007.htm
4/08	Saeed gives an interview to *The News* of Pakistan	http://www.southasianoutlook.com/issues/2008/may/southasia_pakistan.html
12/08	Cites *LeT* publication of several periodicals	http://www.cfr.org/pakistan/lashkar-e-taiba-army-pure-aka-lashkar-e-tayyiba-lashkar-e-toiba-lashkar—taiba/p17882
1/09	Refers to Pakistani government proscribing several LeT publications LongWarJournal cites reports that LeT publications remain available	http://www.jamestown.org/single/?no_cache=1&tx_ttnews%5Bswords%5D=8fd5893941d69d0be3f378576261ae3e&tx_ttnews%5Bany_of_the_words%5D=Nanhay&tx_ttnews%5Btt_news%5D=34421&tx_ttnews%5BbackPid%5D=7&cHash=1e5e7929618ee1cc202dd38ca4dc2233 http://www.longwarjournal.org/archives/2009/01/banned_pakistani_ter.php
3/09	LeT statement to *AFP*	http://www.google.com/hostednews/afp/article/ALeqM5iXFrqCIYMnjS46dmXnB3uzUkNQmQ
7/09		http://www.risingkashmir.com/?option=com_content&task=view&id=14538 (Link broken)
8/09		http://www.sananews.com.pk/english/2009/08/11/let-extends-support-to-aug-11-strike-call/ (Link broken)
9/09	Reference to LeT publication *Ghazva*	http://articles.timesofindia.indiatimes.com/2009-09-25/pakistan/28086357_1_jud-activists-jihadi-hafiz-saeed
11/09	LeT publications appearing under new titles	http://ibnlive.in.com/news/a-year-on-thank-god-we-are-fine-says-jud-leader/105767-2.html?from=tn
12/09	Letter by Saeed published in *The News*	http://www.memri.org/report/en/0/0/0/0/0/0/4074.htm
2/10	Reference to LeT publications	http://www.hindustantimes.com/News-Feed/India/Everything-you-want-to-know-about-the-LeT/Article1-511059.aspx
5/10	LeT spokesperson makes a statement to the media	http://www.thaindian.com/newsportal/south-asia/why-name-saeed-lakhvi-in-2611-verdict-asks-jamaat-ud-dawa_100358062.html
9/10	Pakistani paper prints Saeed statement on al-Qaeda	http://pakistanmediawatch.com/tag/daily-khabrian/

Appendix H
Some LeT Locations Including Training Camps

V. S. Subrahmanian et al., *Computational Analysis of Terrorist Groups: Lashkar-e-Taiba*,
DOI: 10.1007/978-1-4614-4769-6, © Springer Science+Business Media New York 2013

Quarter	Region	Source
*Khanewal office	Punjab	http://www.satp.org/satporgtp/countries/india/states/jandk/terrorist_outfits/lashkar_e_toiba_lt.htm
?	Afghanistan	www.nefafoundation.org/miscellaneous/ekletwitnessreport.pdf
[Seminary I]	Sindh	
[Seminary II]	Sindh	
Abbottabad	NWFP	http://www.satp.org/satporgtp/countries/india/states/jandk/terrorist_outfits/lashkar_e_toiba_lt.htm
Abdul-Bin-Masud camp	PoK	http://www.thaindian.com/newsportal/health/terrorist-training-camps-still-intact-in-pok-say-its-leaders_10046123.html
Abdullah bin Massoud	PoK	www.nefafoundation.org/miscellaneous/ekletwitnessreport.pdf
Al Dawa Model School	NWFP	http://www.satp.org/satporgtp/countries/india/states/jandk/terrorist_outfits/lashkar_e_toiba_lt.htm
Al-Massada	PoK	www.nefafoundation.org/miscellaneous/ekletwitnessreport.pdf
Aqsa camp	PoK	http://www.kashmirlive.com/story/Baglihar-Dam-on-LeT-radar-say-arrested-militants/488507.html
Arifwala office	Punjab	http://www.satp.org/satporgtp/countries/india/states/jandk/terrorist_outfits/lashkar_e_toiba_lt.htm
Athmuqam	PoK	http://www.hindu.com/2009/04/02/stories/2009040255371000.htm
Azizabad	Sindh	http://news.in.msn.com/national/article.aspx?cp-documentid=1715745
Badani Nallah	J&K	http://www.khabrein.info/index.php?option=com_content&task=view&id=19521&Itemid=88
Badli camp	PoK	http://www.thaindian.com/newsportal/health/terrorist-training-camps-still-intact-in-pok-say-its-leaders_10046123.html
Bahawalnagar	Punjab	http://www.satp.org/satporgtp/countries/india/states/jandk/terrorist_outfits/lashkar_e_toiba_lt.htm
Bait-ul-Mujahideen	PoK/ Pak	http://www.hindustantimes.com/StoryPage/FullcoverageStoryPage.aspx?sectionName=&id=9eee80ac-5535-4700-a2e8-9cb6d1f382edMumbaiunderattack_Special&&Headline=Lakhvi+also+behind+Mumbai+train+blasts
Bajaur	NWFP	http://www.satp.org/satporgtp/countries/india/states/jandk/terrorist_outfits/lashkar_e_toiba_lt.htm
Benazir Bhutto Rd		http://www.satp.org/satporgtp/countries/india/states/jandk/terrorist_outfits/lashkar_e_toiba_lt.htm
Bhatta Chowk	Punjab	http://www.hindustantimes.com/StoryPage/FullcoverageStoryPage.aspx?sectionName=HomePage&id=9eee80ac-5535-4700-a2e8-9cb6d1f382edMumbaiunderattack_Special&&Headline=Lakhvi+also+behind+Mumbai+train+blasts
Bhawalpur office	Punjab	http://www.satp.org/satporgtp/countries/india/states/jandk/terrorist_outfits/lashkar_e_toiba_lt.htm
Camp Tango	Afghanistan	www.nefafoundation.org/miscellaneous/ekletwitnessreport.pdf
Chakdara	Lower Dir	http://www.satp.org/satporgtp/countries/india/states/jandk/terrorist_outfits/lashkar_e_toiba_lt.htm

(continued)

(continued)

Quarter	Region	Source
Chamberlain Rd	Punjab	http://www.dawn.com/2008/12/12/top1.htm
Cherapadi Pahadi	PoK	http://news.in.msn.com/national/article.aspx?cp-documentid=1715743
Chowk Chuberzi	Punjab	http://www.hindu.com/2006/08/15/stories/2006081518040300.htm
Circular Rd	Punjab	http://www.dawn.com/2008/12/12/top1.htm
Danna camp	PoK	http://www.thaindian.com/newsportal/health/terrorist-training-camps-still-intact-in-pok-say-its-leaders_10046123.html
Dori	Sindh, Pakistan	http://ibnlive.in.com/news/2611-attackers-training-ground-unearthed-in-pak/85388-3.html
Dulai area	PoK	http://blog.taragana.com/n/jud-expanding-operations-recruiting-in-pok-96302/
Faisalabad	Punjab	http://www.state.gov/s/ct/rls/crt/2007/103714.htm
Fowara/ Fawara Chowk	NWFP	http://www.dawn.com/2008/12/12/top1.htm
G-6/4, street 35		http://www.satp.org/satporgtp/countries/india/states/jandk/terrorist_outfits/lashkar_e_toiba_lt.htm
Gulshan-e-Iqbal	Sindh	http://www.dawn.com/2008/12/12/top1.htm
Haripur	NWFP	http://www.satp.org/satporgtp/countries/india/states/jandk/terrorist_outfits/lashkar_e_toiba_lt.htm
I-8 Markaz		http://www.satp.org/satporgtp/countries/india/states/jandk/terrorist_outfits/lashkar_e_toiba_lt.htm
Ibn Taimiyya	PoK	www.nefafoundation.org/miscellaneous/ekletwitnessreport.pdf
Jamia Qudsia Mosque	Punjab	http://www.dawn.com/2008/12/12/top1.htm
Johar town	Punjab	http://www.dawn.com/2008/12/12/top1.htm
Kahuta center		http://www.globalsecurity.org/military/world/war/kashmir-2002.htm
Karakoram hwy	NWFP	http://www.dawn.com/2008/12/12/top1.htm
Kashmari bazaar		http://www.satp.org/satporgtp/countries/india/states/jandk/terrorist_outfits/lashkar_e_toiba_lt.htm
Kel		http://www.hindu.com/2009/04/02/stories/2009040255520100.htm
Korri road		http://www.satp.org/satporgtp/countries/india/states/jandk/terrorist_outfits/lashkar_e_toiba_lt.htm
Malakand	NWFP	http://www.satp.org/satporgtp/countries/india/states/jandk/terrorist_outfits/lashkar_e_toiba_lt.htm
Mardan	NWFP	http://www.satp.org/satporgtp/countries/india/states/jandk/terrorist_outfits/lashkar_e_toiba_lt.htm
Markaz-e-Hafsa	NWFP	http://www.satp.org/satporgtp/countries/india/states/jandk/terrorist_outfits/lashkar_e_toiba_lt.htm
Markaz-e-Taiba	Punjab	http://en.wikipedia.org/wiki/Lashkar_E_Tayyiba_training_camp
Markaz-e-Taiba	Punjab	http://www.dawn.com/2008/12/12/top1.htm
Markaz-e-Taiba	Punjab	http://www.satp.org/satporgtp/countries/india/states/jandk/terrorist_outfits/lashkar_e_toiba_lt.htm

(continued)

(continued)

Quarter	Region	Source
Markaz-e-Taiyyaba	PoK/ Punjab	http://news.in.msn.com/national/article.aspx?cp-documentid=1715744
Marwah	Punjab	http://www.hindustantimes.com/StoryPage/FullcoverageStoryPage.aspx?sectionName=&id=9eee80ac-5535-4700-a2e8-9cb6d1f382edMumbaiunderattack_Special&&Headline=Lakhvi+also+behind+Mumbai+train+blasts
		www.nefafoundation.org/miscellaneous/ekletwitnessreport.pdf
Moaskar al-Aqsa	PoK	http://www.dawn.com/2008/12/12/top1.htm
Moch Darwaz	Punjab	http://timesofindia.indiatimes.com/In_US_LeT_chiefs_kin_raised_funds_for_jihad/articleshow/3964960.cms
Moon Chowk	Punjab	http://www.rediff.com/news/2008/dec/18mumterror-how-lashkars-terrorists-are-tech-savvy.htm?zcc=rl
Muzaffarabad	PoK	http://www.youtube.com/watch?v=qxHAbPTeoBA
Muzaffarabad	PoK	http://www.satp.org/satporgtp/countries/india/states/jandk/terrorist_outfits/lashkar_e_toiba_lt.htm
Muzaffarabad	PoK	http://www.jamestown.org/fileadmin/JamestownContent/Book_Images/TM_007_1.pdf
Nankana Sahib	Punjab, West	http://www.satp.org/satporgtp/countries/india/states/jandk/terrorist_outfits/lashkar_e_toiba_lt.htm
Neelum valley	*PoK	http://www.dawn.com/2008/12/12/top1.htm
New Zarghoon Rd	Balochistan	http://www.thestatesman.net/page.news.php?clid=2&theme=&usrsess=1&id=238911
Parhana	Punjab	http://www.satp.org/satporgtp/countries/india/states/jandk/terrorist_outfits/lashkar_e_toiba_lt.htm
Pati Tar	Poonch	http://www.asianews.com.pk/lashkar-e-toiba-hideout-in-doda-26995
Pindora	J&K	http://www.dawn.com/2008/12/12/top1.htm
Puneja	Punjab	http://www.dawn.com/2008/12/12/top1.htm
Punjab-presences	Balochistan	http://www.satp.org/satporgtp/countries/india/states/jandk/terrorist_outfits/lashkar_e_toiba_lt.htm
Quetta office		http://www.satp.org/satporgtp/countries/india/states/jandk/terrorist_outfits/lashkar_e_toiba_lt.htm
Rahim Yar Khan office	Punjab	http://www.hindu.com/2006/08/15/stories/2006081518040300.htm
Rajanpur	Punjab	http://www.satp.org/satporgtp/countries/india/states/jandk/terrorist_outfits/lashkar_e_toiba_lt.htm
Rajouri-camp	PoK	http://www.crisisgroup.org/home/index.cfm?id=6010&l=1
Rasheedabad Chowk	Punjab	http://www.satp.org/satporgtp/countries/india/states/jandk/terrorist_outfits/lashkar_e_toiba_lt.htm
Rawalpindi/ Islamabad		http://www.longwarjournal.com/archives/2008/12/pakistan_detains_las.php
Satellite town		http://uk.reuters.com/article/topNews/idUKTRE4B00CW20081207?rpc=401&
Sh(a)wai Nullah	PoK	http://www.sindhtoday.net/south-asia/96427.htm
Shawai Nullah	PoK	http://www.hindu.com/2006/08/15/stories/2006081518040300.htm
Sher Gur	NWFP	
Shiwai Nala		

(continued)

(continued)

Quarter	Region	Source
Sialkot	Punjab, Pak	http://www.dawn.com/2008/12/12/top1.htm
Sialkot presence	Punjab	http://www.indianexpress.com/news/six-lashkar-men-who-trained-with-mumbai-attackers-holed-up-in-kashmir/408594/
Sindh presences	Sindh	
Swabi	NWFP	http://www.satp.org/satporgtp/countries/india/states/jandk/terrorist_outfits/lashkar_e_toiba_lt.htm
Tarambadi Chowk		http://www.hindu.com/2006/08/15/stories/2006081518040300.htm
Tareen Rd	Punjab	http://www.satp.org/satporgtp/countries/india/states/jandk/terrorist_outfits/lashkar_e_toiba_lt.htm
Thatha training camp	Sindh	http://www.tmcnet.com/usubmit/2009/07/26/4291861.htm
Tench Bhatta		http://www.satp.org/satporgtp/countries/india/states/jandk/terrorist_outfits/lashkar_e_toiba_lt.htm
Timergara	NWFP	http://www.hindu.com/2009/04/11/stories/2009041156041000.htm
Zaffarwal center		http://www.globalsecurity.org/military/world/war/kashmir-2002.htm

References

Ernst, J., & Subrahmanian, V. S. (2004). *Method and System for Optimal Data Diagnosis*, US Patent 7474987, issued Jan. 6, 2009. Filed Feb 10, 2004.

Khuller, S., Martinez, V., Nau, D., Simari, G., Sliva, A., & Subrahmanian, V. S. (2007). Computing most probable worlds of action probabilistic logic programs: Scalable estimation for 1030,000 worlds. *Annals of Mathematics and Artificial Intelligence, 51*(2–4), 295–331.

Mannes, A., Michael, M., Pate, A., Sliva, A., Subrahmanian, V. S., & Wilkenfeld, J. (2008a). Stochastic opponent modelling agents: A case study with Hezbollah.In H. Liu, J. Salerno, & M. Rogers (Eds.), *Proceedings of the 2008 First International Workshop on Social Computing, Behavioral Modeling and Prediction*, Springer, Phoenix, April 1–2, 2008.

Mannes, A., Sliva, A., Subrahmanian, V. S., Wilkenfeld, J. (2008b). Stochastic opponent modeling agents: A case study with Hamas. In *Proceedings of the 2008 International Conference on Computational Cultural Dynamics*, (pp. 49–54). September 2008, AAAI Press.

Parker, A., Simari, G. I., Sliva, A., & Subrahmanian, V. S. (2011). Approximate achievability in event databases. *Proceedings of the ECSQARU, 2011*, 737–774.

Simari, G. I., & Subrahmanian, V. S. (2010). *Abductive Inference in Probabilistic Logic Programs, Technical Communications of the 2010 International Conference on Logic Programming*, (pp. 192–220). Edinburgh, UK, July 2010.

V. S. Subrahmanian et al., *Computational Analysis of Terrorist Groups: Lashkar-e-Taiba*, 225
DOI: 10.1007/978-1-4614-4769-6, © Springer Science+Business Media New York 2013

Index

V. S. Subrahmanian et al., *Computational Analysis of Terrorist Groups: Lashkar-e-Taiba*,   227
DOI: 10.1007/978-1-4614-4769-6, © Springer Science+Business Media New York 2013